Walking in a Scientist's Shoes

The People You Will Meet and the Places You Will Go

By

A Martin Gerdes, PhD

Acknowledgements

There are so many people to thank: My parents and grandparents for their kindness and love; my brother Chris, and my sisters, Nancy and Delores (bless her soul, she passed from Alzheimer disease a few years ago); all the wonderful women who put up with my bullshit and danced with me (and please forgive me for all my sins). There are so many amazing people who passed through my scientific world: students, postdocs, technicians, fellow faculty members and scientists too numerous to count. Many have been mentioned throughout the book but there are so many more. I would also like to apologize again to the retired Associate Vice Chancellor of the University of Illinois Chicago, a fellow heart researcher, for repeatedly stepping on her toes at that meeting a few years ago while I was trying to remember how to do the Waltz. The best way to avoid leaving anyone out is to provide a search link for my publications since I will be eternally grateful for the hard work of each co-author and those mentioned here.

https://www.ncbi.nlm.nih.gov/sites/myncbi/1pynPyuqrF0Af/bibliography/40151682/public/?sort=date&direction=descending

Table of Contents

About the Author

Anthony Martin Gerdes grew up in Edna, a small Texas town of about 6,000 people located midway between Houston and Corpus Christi. It is a farming-ranching community. His dad owned a laundry-dry cleaning shop. Martin became interested in photography and built a darkroom in his bedroom in the 8th grade. He started a professional wedding photography business in high school which enabled him to buy his own car and pay his way though college. Having a strong interest in science, he eventually obtained a PhD from the University of Texas Medical Branch at Galveston. An idea in graduate school led to a life-long quest to discover an effective treatment for heart failure. It was a very meaningful and rewarding career. Martin also grew up with a passion for music and dancing. The book is filled with many amusing stories of a life that has been far from boring.

Martin has been affiliated with 8 different medical schools, has published over 130 scientific papers, obtained over $30M in research funding, and won international awards for his research. In his later career, Martin built three successful research programs as a department chair or research center director. Along the way, he became friends with some of the leading minds in medicine and was a major contributor to several important advances in cardiology.

Overview

Most people know science only by its end results: a new drug, a new guideline, a headline announcing a "breakthrough." What they almost never see is the long, uncertain, often messy human journey that leads there, the false starts, the stubborn ideas that refuse to go away, the failures that teach more than successes, and the individuals whose curiosity quietly reshapes medicine over decades rather than overnight.

Walking in a Scientist's Shoes invites readers inside that hidden world. This book is a narrative journey through one life in biomedical science, beginning in a small Texas town with little guidance toward higher education, and unfolding into a career that moved from graduate-student curiosity to discoveries with the likelihood of changing how heart failure is treated. Along the way, it demystifies how science actually works, not as a straight line from idea to cure, but as a deeply human process driven by persistence, imagination, and collaboration.

Written for **educated non-scientists**, the book explains complex ideas, especially heart failure and cardiovascular research, in clear, accessible language, often with humor, always with empathy. Readers do not need a scientific background to follow the story; they simply need curiosity about how knowledge is built and why it matters.

A Story of Becoming

The first part of the book traces the formative years of a scientist who did not grow up surrounded by mentors, blueprints, or assurances. Growing up in a small Texas town, the path toward science was neither obvious nor encouraged. Instead, it emerged gradually, from curiosity, from work, from watching how people solve problems with limited resources, and developing the ability to recognize and pounce on real opportunities while ignoring the bright shiny objects that distract.

These early experiences shaped an unconventional scientific instinct: a reluctance to follow bandwagons, a sensitivity to overlooked ideas, and a belief that insight is not confined to credentials. For readers, this section offers reassurance that science is not reserved for prodigies or insiders, but for people who ask questions and refuse to let go of them.

How Discoveries Really Happen

The heart of the book follows the evolution of a single idea, from a graduate student's hypothesis to a body of work that now appears to represent a major advance in heart failure treatment. My hypothesis was: **Understanding structural remodeling of the heart in progression to failure is the key to developing new treatments that enable patients to live longer**. Guess what? It looks like I was right! A major clinical trial is now underway at the National Institutes of Health (NIH) to investigate a new treatment that is the focus of my life's work. Rather than presenting research as a technical achievement, the book tells the story of discovery as lived experience: late-night excitement as new ideas suddenly come to light, some stretches of doubt, and moments when data suddenly force a re-thinking of what is possible.

Heart failure, one of the most common and devastating conditions worldwide, is explained in plain language, with metaphors and stories that allow readers to understand what goes wrong, why treatment is difficult, and why progress has been so slow. Against this backdrop, the book raises a provocative question: **What if a highly effective, low-cost treatment for heart failure has been hiding in plain sight?** And why might such an idea be neglected for so many years? Rather than offering simple answers, the book shows how science advances in the real world, through repeated testing, skepticism, institutional constraints, and the courage to continue even when other experts fail to recognize the importance of the work.

Inside Academic Medicine and Leadership

In its final section, **Walking in a Scientist's Shoes** pulls back the curtain on academic medicine itself. As a department chair and director of a cardiovascular research center, the author built three major research programs at different institutions, each under different conditions, constraints, and cultures. Through story rather than instruction, readers come to understand how medical schools function, how leadership decisions shape scientific success or failure, and why great ideas can flourish, or wither, depending on the environment that surrounds them.

This is not a manual on leadership. Instead, it is a candid account of what strong and weak leadership looks like in practice, how influence works, and how persistence extends beyond the laboratory into institutions and policy. Along the way, readers see how discovery is not only a scientific act, but an organizational one.

Why This Book, Why Now

At a time when public trust in science is fragile and the pathway into scientific careers feels opaque or discouraging, **Walking in a Scientist's Shoes** offers something rare: a grounded, personal account of what a life in science truly looks like. It challenges the perception that science is boring, inaccessible, or impersonal, and replaces it with a picture that is unpredictable, demanding, occasionally frustrating, and often exhilarating.

The book also addresses a widening gap between the public and the scientific enterprise. By showing how discoveries mature over decades rather than news cycles, and by humanizing the people behind them, it helps readers understand not only *what* science produces, but *why* patience, funding, and persistence matter.

Humor threads throughout, stories of coming of age, living through societal transformation during the 1970s including the Vietnam war, interesting colleagues, unexpected experiences, and moments of levity that remind readers that scientists are not abstractions or stereotypes, but people who may be more like you than you think while different at the same time.

A Book for Curious Readers

Walking in a Scientist's Shoes is ultimately an invitation to step inside the daily life of a scientist, to see how ideas take root, and to appreciate the quiet determination behind medical progress. The author also goes through his progression from childhood to graduate school and many, often funny and sometimes scary, life experiences that many people can relate to. Young readers considering science will see that the path is not predetermined. Scientists can come from anywhere. Readers will gain a deeper understanding of how discoveries that affect them and their families come to be. By the final page, readers may not want to become scientists themselves, but they will understand why some people cannot imagine being anything else.

PART ONE

Personal Journey and Foundation

Chapter One
Introduction

I grew up in a small Texas town with male classmates who would typically go on to become farmers, ranchers, construction workers, plumbers, etc. and female classmates who generally thought that teaching and nursing were their only career options. From an early age I loved science. I had this nagging feeling that I was put on this planet to do something important and was pretty sure it had something to do with science. But I had no idea what that was or how to pursue it. My town was filled with hard-working people who raise our food, provide our beef, and fight our wars. But no one had a clue about how to become a scientist. Surprisingly, my classmates apparently knew what I would become before I did. I earned the nickname "Professor" before entering high school. I had so many interests in my early years that I needed to reach that conclusion by process of elimination, with the decision finally culminating in my senior year of college. Thankfully, a brilliant black man came into my life to provide needed mentoring. Mr. Cranford was my high school biology/physics teacher. He was among the few black teachers hired at Edna High School after closing the black high school when segregation was ruled illegal. I will be forever grateful that Mr. Cranford opened my eyes to the world, the challenges, the opportunities, and most of all, the absolute need to be persistent and focused.

I hope my story resonates with the public, especially bright young people considering science careers, and illuminates the path forward. For general readers, I offer an insider's view of a scientist's life, unique, unpredictable, and deeply rewarding. Although the book is largely divided into three parts: my personal story, my scientific story, and building programs in a leadership role, it is mostly progression through my life. It is hard to separate work and life since they happen at the same time and are too intertwined to separate completely.

Let me be honest, the traditional academic path can be challenging. Eight to ten years earning your PhD. Two or more years in postdoctoral training. Competing for faculty positions. Constantly chasing grant funding to keep your job. I took that path. I'd take it again in a heartbeat. But academia isn't the only option. PhDs thrive in biotech, pharmaceuticals, device companies, and countless other fields.

The key is finding work that you love. I have often heard scientists say: "I can't believe I get paid to do this." What do they mean by that? You get into science because you love it. The sky is the limit. You search for a research topic that drives your passion. This takes you down a path of discovery that can feel like you just met the girl of your dreams when you finally find THE project. Basic scientists at medical schools often have about 75% of their time pursuing research, the ideas that pop into their heads. For the most part, YOU choose what you do and pursue ideas that can make you tingle with excitement on the good days. On several occasions, I have gone to work at 3 or 4 AM because I was too excited to sleep. Wouldn't you like to feel like that about your job? The rest of your time as a professor is mostly spent teaching amazing young people who keep you young at heart. Every day is spent surrounded by intelligent, creative people who keep your mind nimble. In some respects, I feel like a 35-year-old trapped in a 70 year old body. But a lifetime of exercise has helped me feel 20 years younger. There is no fountain of youth, but those who stay fit live longer and feel younger.

I want to give readers an idea of how research evolves from an idea and progresses step by step toward an important goal. Along the way, other discoveries may arise that trigger more projects and new directions. Your lab grows and before you know it, there are 10 or more people working in the lab on half a dozen projects. Trainees come to your office with exciting new data with increasing frequency. Because I run a translational research lab, data are often of obvious clinical relevance. You sometimes think: "If this turns out to be true in patients, it will change medicine." That's not boring. Importantly, the type of research I do in relevant animal models almost always relates to the human situation. Discoveries can also lead to improvements in animal care too, helping your pets live longer, healthier lives.

I tell the story of how my work eventually evolved from an idea as a graduate student into what should very likely be one of the most important breakthroughs in cardiology. But it must be confirmed in people and that takes time. This is how you feel when you know in your heart that you have discovered a fundamental truth about the heart. What if there's a highly effective, low-cost treatment for heart failure that is being ignored? How did I come to know that? I will walk you through it. An unusual aspect of this treatment is that it will have NO adverse side-effects when optimally dosed, and it is ready for clinical trials now. Readers will also get a taste of the politics in academia as I tell the story of my time as a medical school department chair and research center director. I would venture to guess that my life has had way more euphoric high points than most people out there working 9-5. I will let you be the judge of that, as you make it through the book. I tried to also relate to the non-scientific public who want to be more knowledgeable about heart diseases affecting themselves, their family, and friends.

Another goal of the book is to share how formative experiences growing up shaped my unique approach to science. It helped me become an innovator rather than just another smart person jumping on the latest bandwagon. I also learned to never underestimate the creativity and insight of people with less formal education. Sadly, I learned that common sense is way too uncommon. I learned the power of persistence. In fact, my favorite dean stated at his retirement dinner that he learned the *definition* of persistence from me. Cumulatively, I believe my early life experiences also resulted in becoming a leader, rather than a follower. Aside from my scientific career, I have successfully built major research success stories at three different places, as a department chair or founding director of a research center. I learned how powerful a well-placed letter to the US Senate Majority Leader can be. Finally, I hope other readers will enjoy the book and have a better understanding that scientists are people just like them (well, almost). If young readers who love science can see themselves walking in my shoes and finding them very comfortable, I will have accomplished my primary goal.

The average person on the street likely has no idea who scientists really are as people. We are more than just a collection of nerds. Some can be among the most interesting people in the world. For instance, my friend Eric Olson, who many consider the most famous heart and skeletal muscle researcher in the world, also has his own rock and roll band where he sings and plays guitar. He co-founded a biotech company, which sold for $2.5B, then started another. Like me, Eric would feel equally comfortable sharing beers and conversation with a third-grade dropout or a Nobel laureate. We might even find the dropout more interesting. He is among the most down-to-earth and humble people I know. Eric is also the Willie Nelson Endowed Chair at the University of Texas Southwestern Medical Center in Dallas. I once asked Eric, "What was the high point of your career?" He had so many important discoveries that I had no idea how he could choose one. His response, "Well, I got to open for Willie once."

Chapter Two
Small Town, Big Dreams: A Texas Beginning

I grew up in a small town in Texas famous for a stripper and a wrestler. I later left Texas and immigrated to the United States, where I became fluent in English. Previously, I spoke Texan, which is a lot like English but without the grammar or punctuation. We all know grammar was invented by Yankees to make us look stupid, right? Seriously, there is some truth in that as I will note later. And, if you read one of my younger sister Nancy's texts, you would learn the meaning of a run on sentence. Someday, I will introduce her to the period and comma, just for starters and not to overwhelm her.

I understand the insecurities typical of growing up in a small town because I had them too. City kids had so many more advantages growing up. The educational inequities seemed most obvious to me. City schools were much larger, had more advanced courses, and had advantages in attracting better teachers because of the attractive local amenities which include major sports, museums, and so many interesting things to do. We had advanced algebra in Edna, but they could not find a qualified teacher. So, the high school football coaches' wife was assigned that task. In our case, I doubt she could even spell algebra. She just assigned a set of problems at the beginning of class and had us go through the teacher's text on our own for answers at the end. We quickly learned it was useless to ask for any further explanation. But we did get lucky from time to time. There was my chemistry teacher, "Tom Cat" Fryer. He was a former rodeo star and had the gaudy belt buckle to prove it, which he proudly wore to school every day in his jeans, white shirt, Bolo necktie, and cowboy hat. He was actually a very good teacher who made chemistry entertaining. His mom had a pet goat that she adored. He asked me to take a small picture of the goat for a broach to give his mom for Christmas. I wondered what he would say when presenting it: "Here you go, you old goat!" Probably not. It's heart-warming that they were so close. I am sure this "Cowboy Chemist" surprised many and was certainly a lesson in why you can't judge a book by its cover.

I first wondered if I could compete against the Houston kids from large high schools a hundred miles away. Then, I wondered how competitive I would be in college. After becoming a scientist, I wondered if I could compete with scientists from top US colleges and from around the world. I was always pleasantly surprised to learn the answer was "yes". But it took a lot of work to get there. Smart kids can come from anywhere and go on to do great things. For instance, shortly after moving to Sioux Falls, SD, later in my career, I learned of the Lawrence brothers from the nearby small town of Canton. Ernest Lawrence (Lawrence-Berkley Labs in CA) would go on to win a Noble Prize in physics for inventing the cyclotron. He started a wave of Nobel Laureates at Stanford while department chair. His brother John is considered by many to be the "Father of nuclear medicine", and some thought he also deserved a Nobel Prize. I bet it would have been interesting listening to conversations between those two as they grew up.

My hometown, Edna, has a population of about 6,000, similar to that of Canton, but without the Nobel Laureate. As indicated earlier, Edna was famous for a stripper and a wrestler, two popular sports in Texas. The stripper was Juanita Slusher, who went by the stage name Candy Barr (for the curious, there are videos of her acts online). She grew up about two blocks from me. Candy Barr (most referred to her by the stage name) was friends with Jack Ruby but worked at a different strip club in Dallas. Apparently, Ruby visited her in Edna before he killed Oswald. I remember hearing that FBI agents visited her in Edna shortly after the assassination due to her relationship with Ruby. I always thought she knew the truth about the Ruby-Oswald shooting and possible Mafia connections. I once asked her, but she just smiled and winked. She went to her grave with that secret. She moved back home while on probation for possession of 1 oz of marijuana, a 15-year sentence. Ironically, she was pardoned by Governor John Connally a few years after the Kennedy assassination. In her heyday, this woman's sexy smile and sultry voice could literally make men melt. My younger brother, Chris, told me the Sexmuseum in Amsterdam had a room dedicated to her. When I mentioned to my dean in South Dakota that she grew up near me, Bob said he and his wife saw her strip in Kansas City. She got around in her day! The wrestler from Edna was "Stone Cold Steve Austin" who went to high school with my brother Chris. Austin later became a big movie star specializing in action movies. Chris speaks highly of Austin, indicating that he was more likely to break up a fight than start one in his early years.

Near the end of my career while living in NY, I noticed with great interest that the New York Times featured two articles about Edna. The first article appeared on December 29, 2012, and featured the Edna Dome. It was a dual-purpose dome for sports and a hurricane shelter, with the article describing a very innovative building, a progressive move by a small, rural town. The second NYT article, featured on the front page, appeared on December 25, 2019 **(Figure 1)**. The article covers the devastation of downtown businesses after Walmart built a store at the edge of town on Highway 59, the major route between Houston and Corpus Christi. More importantly, it wrote about the resilience of downtown Edna after Walmart moved away many years later. Of note, the Edna Theatre was a very important part of my childhood. It was packed for Saturday and Sunday afternoon matinees featuring the latest Hollywood movies, with some of the older high school kids in the upstairs balcony smooching. A block away from the town center were two drugstores with soda fountains. I remember having many cherry Dr Pepper's or ice cream floats as a kid. The only Jewish family in town also had a department store (RBs) where we bought our sneakers, blue jeans, and shirts. All are closed now, unfortunately. But there are nice boutiques, Mexican restaurants, and Bar-B-Q places in Edna. The town is still a slice of Americana with farming and ranching common occupations.

Edna was founded in the late 1800s with the establishment of a railroad depot. An Italian nobleman and investor, Count Telferner, built the railroad through this area. He had three daughters named Inez, Edna, and Louise and named depots after them. There is an old joke about the traveling salesman who slept between Inez and Louise. He spent the night in Edna. Nearly all the residents in nearby Texana moved to Edna when the railroad bypassed the town. It eventually became a ghost town. In 1968, Edna got national attention by proclaiming itself "Flag City USA."

This was a move to show strong support for the government and disdain for the Vietnam anti-war movement. Flags were everywhere with the pledge they would remain until we reached an honorable peace in Vietnam. I believe most are still up.

Figure 1. New York Times article about Edna's resiliency after losing its downtown businesses when Walmart moved in and later left. Dec 25, 2019.

Neither of my parents went to college. My Dad (Tony) was the only one of 11 kids to graduate from high school. That was common in those days, particularly in farming communities. During World War two, he was selected for radar school, since he was the smart one, the only one with a high school diploma. The town my mother grew up in was so small that she skipped the second grade because she would have been the only one in the class. They got married a few months after she graduated from high school. I have included a picture of them when dating and my Mom's senior high school picture **(Figure 2)**. Leverette, a confusing name invented by my grandmother, looked forward to leaving home and a life of partying. I say confusing because she spelled it three different ways and often seemed unsure of the spelling. I think she just found it amusing to increase the number of double letters in her name (Leverettee was another spelling). She turned 17 in July, was married the next month, and had my sister Delores 9 months later, followed by me 14 months later. In the case of Delores, some of her friends were betting that she would not make it to the nine-month goal of respectability. I was born on my mother's birthday. Happy birthday Mom and so much for partying! My father was 11 years older. Most of my siblings and cousins graduated from college, with a very high percentage going into teaching as a profession. Several of us, including me, obtained PhDs and are now university professors. We were mostly the first-generation in our family attending college, apart from my mother's younger brother and sister.

Figure 2. Mom and Dad when dating.

Mom's senior picture

At the suggestion of a friend, my dad bought a dry-cleaning store after leaving the Army and later added laundry and carpet cleaning to his business. He knew nothing about the dry-cleaning business but quickly learned on the fly. As a kid, I often watched him repair things on Sundays after the pipes had cooled off from the boiler being shut down at noon on Saturday. Like me, he had a lot of energy, always busy, looking for things to do or fix. He was a great problem solver and walked me through the process as a kid. I learned a lot from him. He was a very sweet and kind man. I remember when we got into trouble my mom would say, "Just wait until your father gets home." But all he wanted to do was hug us. So, spankings were few and far between.

One of my earliest childhood memories involved my mom going to the hospital. My Dad was beside himself and crying when he told us. It turned out that she had a nervous breakdown (depression) due to the overwhelming responsibilities of having two babies by the time she was 19 and who knows how many more on the way. Like the men of the "Mad Men" era, my dad enjoyed meeting his buddies after work for a few beers or participating on a weekly bowling league, with the little lady waiting for him at home with a wonderful evening meal. He discussed this with me later after I grew up. Mom's condition was really a wake-up call. He thought he had lost the love of his life. Back then, mental health awareness was underappreciated. After that, he started hiring a baby-sitter and taking Mom out dancing more often. It did not take much to make her happy.

By the time my sister Delores and I were about 10 and 12, my parents let us babysit with Nancy. There was a tree growing next to the house in back and Delores and I liked to climb onto the roof and play. One time, we tied a rope around Nancy and pulled her up to the roof too. She was about 2 as I remember. We never told our parents about this until adults. We did not want to lose the babysitting money. I don't think it ever dawned on us that it might be a bit dangerous. I remember jumping off the roof once to see what it felt like. Luckily it was a ranch house and I was sitting on the edge of the roof instead of taking a running start. Nonetheless, it took my lower extremity a few days to feel right again. In the meantime, I guess I proved a well-known hypothesis, smart people can do stupid things. That was probably my first experiment and the results were so clear that a repeat was unnecessary. Sometimes an "n" of 1 works.

My mother said I was good at fractions at an early age. For instance, Mom said I once sat on the road in front of our house at about age 3. A lady stopped her car in the road and said: "Son you need to get out of the road." I told her: "I'm not your son. It's our half of the road, in from of our house, so I did not need to move." She took me inside and my mother made me promise to never do that again. I have no memory of that but Mom loved telling the story after I grew up. Unlike today, kids were on their own at an early age back then.

The other dry cleaner in town was a bit of a racist who went out of business and sold to my dad. Let's see? Who dresses up in their best for church on Sunday? I remember when waiting on customers at my dad's cleaners on Saturday afternoon that black customers were often so poor that they "split bags" (took only part of the order), so they had a nice outfit for Sunday. Sunday church was a celebration you dressed up for. In contrast, I noticed many white boys wearing dirty blue jeans and T-shirts to our Catholic church on Sundays. So, in my dad's case treating everyone with respect and dignity regardless of skin color turned out to be more profitable. We were well-dressed on Sunday. As the owner of a dry-cleaning store, it was a requirement.

I enjoyed getting to know my dad's employees, who were mostly black and Hispanic. A Mexican lady, Cookie, was a huge flirt and funny as hell. She was married but there was always a parade of guys dropping by to flirt with her. A black man, Buster Gibson, had the reputation of being an amazing presser at the cleaners, never a double pleat. He once dropped a coin on the floor and asked me to pick it up. I blushed when I saw the image of a naked woman on the coin. He liked to kid around with me. I really enjoyed his playful spirit. But he was an alcoholic and my dad would often have to go awaken and sober him up for work. One chilly morning, Buster did not show up for work. My Dad learned that he had passed out in his front yard and froze to death (yes, on the Texas Gulf coast). It was clear that my dad, who did his best to save Buster, believed he failed. He could not keep the tears back when he told us.

My neighbor Randy once talked me into stealing strawberries from our neighbor's patch. We were about five years old. We snuck over to the patch, grabbed a few strawberries, washed and ate them. Strawberries grow about ankle height. So, we must have looked ridiculous crouching down and then quickly running back home with 3 or 4 strawberries each. That afternoon my mom asked me, "Did you two boys steal strawberries from Mrs. Curley's patch? I responded, "If I tell you, will you whip me?" Thankfully, my reasoning skills have matured a bit since then and I later learned a new word, incriminating.

My hometown was half white and about a quarter Hispanic and a quarter black. The town was highly segregated with few whites venturing into "colored town". We would often go for a Sunday afternoon drive, mostly visiting relatives in town or at their nearby farms. But, on many occasions we would end up at the house of Lloyd and Itasca Stafford, the most highly respected black family in town. Itasca was a high school English teacher, one of the few black teachers hired when the black and white schools integrated about the time I entered high school. Her husband, Lloyd, owned a taxi service and motel. We would spend hours over there visiting with them and their many friends who dropped by. I suspect we were the only white family in town who did something like that.

When I was a kid, I remember a big disturbance regarding the deputy sheriff tying a black man to a tree and horse whipping him. While sharing a beer with a black man I grew up with many years later, he told me the rest of the story. Apparently, the deputy sheriff was having an affair with the black man's wife (not clear if it was consensual). The deputy sheriff had a beautiful teen age daughter and was told the black man had flirted with her. I never confirmed that part of the story, but it made sense. There was a lot of racial tension back then.

We did not have a public swimming pool in Edna. The Edna Motel made additional revenue by charging locals to use their pool. But you had to be white. I remember some of my white childhood friends actually thought the water would turn black if they were allowed to swim there. After segregation became illegal, they chose to close the pool rather than admit blacks. So, we would often drive 8 miles to Ganado where there was a very nice public pool. They responded to the integration order, but there were few minorities there. I doubt many blacks in town could swim in those days since there were no opportunities for lessons. Some of the most memorable songs from my youth were linked to that pool in Ganado while watching everyone grow up…"Cherish, Baby Love, Can't take my eyes off you, Surfer Girl, She loves you, Baby it's you, Respect, Light my fire, Good vibrations, Proud Mary, Under the boardwalk, Wild thing…"

Amazingly, it turns out that the best teacher I ever had in my life was a black man, Evies Cranford, who taught me biology and physics in high school. It was rather easy for me to graduate at the top of my class of about 130. But he pushed me hard, often reminding me that my competition was not in Edna. It was sheer luck that Evies ended up in my town. His wife grew up there. He was president of the student body his junior and senior year at the **HBCU (Historically Black Colleges and Universities)** he attended. His humor encouraged the class clowns to pay attention, or they would miss many jokes, particularly in Biology class where reproduction was a necessary discussion topic. I believe a bit of his style of humor rubbed off on me. He was an incredible teacher with an amazing personality. He later got a PhD and became a high school administrator in Kansas. From him, I learned that it was not only important for young black boys to have a role model, but also for white kids growing up in segregated neighborhoods to learn that there are some very intelligent black people out there who are worthy of your respect. Evies was always the smartest guy in the room. I sometimes noted this made his mostly white peers a bit uncomfortable. I was not surprised that he eventually divorced, moved away, and married another woman with a PhD. I managed to visit them a few times as an adult and frequently reminisce about those wonderful encounters.

My other favorite teacher was Domingo Toledo, who taught Spanish. He was such a sweet man and a very good teacher. The students loved to tease him though, because of his Puerto Rican accent. Me and a friend convinced him the classroom needed a Christmas tree. So, he let us out of class to go drive around and find one. We ended up at the cemetery and cut off the side of a nice cedar tree, without getting caught. We told Mr. Toledo it was a very special corner tree since about 2/3 of the tree was bare. The next few days, classmates religiously decorated the tree. They walked very slowly in procession to kill as much time as possible. Kids in the class would often impersonate him, mostly behind his back. "*I tella you ten or four times to charup. You no charup, I send you principals orifice.*" He once said. A student behind me said, "Oh no, not that smelly thing." He did not get the joke, and I was not about to explain. I sometimes saw Mr. Toledo working at a convenience store in nearby Victoria on weekends to fully support his family. It is so sad that we don't pay these wonderful people a living wage to teach our most important asset, our children.

My high school English teacher was Ms. Paine. She was demanding and ran a very tight ship. Although I was determined to make English my least favorite subject, I still made A's and learned a lot. She made us read the "Owl and the pussycat" out loud. No one could get through that poem without laughing, and the guys were often quoting their favorite line for the next few weeks. I was surprised when she attended a talk I gave in Edna at my parents' 50th anniversary. She was in her early 90s and drove over 100 miles to attend. It was so much fun to catch up with her as an adult. It was surprising to find that I was one of her favorite students and she loved following my career in the Edna Herald, our weekly newspaper. I had no idea. She was all business.

I attended St. Agnes Catholic School through the 6th grade and was an altar boy until starting high school, when I did a stint as a lay reader. I became good friends with Father Flannigan. When visiting home during my first few years of college, I would sometimes drop by the rectory and discuss philosophical questions related to religion. Over time, I was becoming rather agnostic but we both enjoyed the discussions, particularly after I had taken a philosophy course in college. After enrolling in that class, I learned that the professor had never given an A, but everyone passed if they showed up. I made the first A he ever gave. I forget the specifics of my philosophical argument on the final exam but remember using elephant's butts to prove the existence of God. I knew I had to be creative if there was any chance of getting an A.

I got a spanking once in Catholic school. We were practicing for first communion in church when the boy in front of me let out a loud fart that resonated throughout the quiet church. Several of us were laughing so hard we could not control ourselves. Sister Lupe marched us back to the school. She retrieved a short 2X4 board from a trash can. It had a huge nail sticking out of one end. She grabbed the board and it appeared that she was going to drive that nail into our butts. But, at the last minute she switched ends and lightly patted us on the rear. I believe we were all terrorized for life. I still have dreams about it.

I was active in the Cub Scouts (mom was a Den Mother) and Boy Scouts. I eventually became an eagle scout with 2 palms. My mother framed my merit badge sash, and it hangs in my office today. I was very much a nerd at that time. But I really enjoyed nature and camping out. I literally built the troop, which had become inactive until I convinced a lay teacher at the Catholic school to become our scoutmaster. There were very few merit badge councilors, so I had to recruit most of them if there was any hope of becoming an eagle scout. I particularly liked first aid and got my first inkling that medicine might be in my future at a regional first aid meet. We were, by far, the youngest troop at the meet. I was a 1st Class Scout at the time but no one else had made it past Tenderfoot, the lowest rank. Edna's biggest town rival was Ganado. Ganado is Spanish for cattle, and we had a lot of them in this area. There were always a few fights at Edna-Ganado football games. Interestingly, we were the Edna Cowboys and they were the Ganado Indians. Ganado had a very mature troop with several Eagle scouts at the time. At the first aid meet, teams of about 6 scouts would pass through different stations where we were confronted with a medical emergency. The leader would organize the team to act out the response. Being the only one in the troop who had passed first aid, I was the leader. The result shocked everyone at the meet. We won 1st place! I have included a picture of some of the boys in Troop 138 **(Figure 3)**. I discovered girls shortly after the picture was taken and left the scouts. In addition to me, several others from my troop later made Eagle rank. Scouting is a wonderful experience that I wish more boys would pursue.

Figure 3. Some of the scouts from Troop 138. I am 2nd from the left.

While in the 8th grade, I became interested in photography and turned my bedroom into a darkroom. Before that, it was a chemistry lab bench with a distilled water setup (I had a gas line for a Bunsen burner) and a microscope. That reminds me of a conversation I once had with my friend Dr. Gary Francis, a famous heart failure expert who was once President of the Heart Failure Society of America and former Director of Cardiac Intensive Care at the Cleveland Clinic. We both had a chemistry set in our bedroom when we were growing up. But it appeared that Gary may have come closer to blowing up the house than me. The conversion of my bedroom chemistry lab to a darkroom probably brought some peace of mind to my parents. From savings earned at Dad's cleaners, I bought a decent camera, an enlarger, and started doing a lot of black and white photography.

When I started high school, it became apparent that I was the only student who knew much about photography. So, I became the school photographer and master of a very nice darkroom at school. A good thing about being school photographer was that me and my assistant could come and go as we pleased. If anyone asked, we would just say we were on a photo assignment. I even had a key to the outside door of the high school so I could do darkroom work on weekends. Thinking back, it's hard to imagine that happened. During my junior and senior years, I enjoyed working with the editor of the school annual, Donna Riddle. It was a point of pride for me since annuals are largely driven by the photos. She was rather progressive for our little town. We created an annual that many found a bit "different" (too progressive), but it aged well as noted from positive feedback at class reunions over the years. Donna eventually went to Hollywood, tried her luck at acting, and later made an award-winning documentary about horses. Apparently, a lot of girls in Edna loved horses. Unfortunately, one girl in my class was kicked in the head by her horse and killed. It wasn't the horses' fault. Surprisingly, her younger sister went on to become the women's high school national rodeo champion while also winning the associated national high school rodeo beauty queen title. These little country towns produce some tough, attractive women.

By the end of my freshman year, I had gained a reputation as a good photographer. The prettiest girl in town, who had just been named Miss Edna, asked me to take her wedding pictures to save money. She was a graduating senior who came from a poor family. Before long, I began to get more weddings. By the time I graduated from high school, I had taken a considerable chunk out of the local wedding photographer's business. I think he was glad to see me leave. But my mom continued to book jobs for me throughout college. I managed to make it through 8 years of college with only $10,000 in college loans. It certainly helped that Texas had one of the lowest tuition rates in the US. The most I ever paid was $130 one semester. It was *so high* because I took 20 hours and 3 lab courses that semester. Rather than my parents spoiling me, I spoiled them. This left more money to spend on my two sisters, Delores and Nancy (8 years younger), and my brother Chris (16 years younger). Apparently, after Delores and I were born, my mother discovered birth control. She later indicated that a high school friend told her about the Rhythm Method, approved by the Catholic Church, but got it backwards. After two quick pregnancies, she sought other advice. I am grateful that she waited.

During my senior year, the deputy sheriff (not the same as mentioned earlier) asked me if I could assist with photography after the passing of the police photographer. To see if I was up to the job, he showed me some photos. It appeared that a man passing through town had accidently hung himself in a motel. His pants were around his legs and a Venetian blind cord, wrapped in a washcloth, was wrapped around his neck. It was ruled an accidental suicide. But it looked like he died happy based on the photographic evidence. Looking at those pictures was my only experience with forensic photography. I made it through my senior year without a phone call from the sheriff's office. Not a lot going on in Edna. It was a good place to get beat up but deaths by accidental hanging were relatively rare.

I have continued to enjoy photography throughout my life and even won a few photo contests at work (New York Institute of Technology). I have not shot any weddings in quite a while. I sometimes wonder if you even need a professional camera with the quality of cell phone cameras these days. This evolution has led to so many people becoming great photographers. Living on the North Shore of beautiful Long Island, I have become particularly fond of sunrises and sunsets over water. I also like to put some of my pictures through a Van Gogh filter to make them a bit more creative. I have included a few photos here **(Figure 4)**.

Figure 4. Examples of some of my recent photography. Closeup of bee and sunflower (top left, grand prize winner at the 2025 NYIT summer photography contest). Sunrise in Oyster Bay Marina (top right). New York City bikes and Empire State building (bottom left, another NYIT award winner). Washington Square in New York City (bottom right).

While in high school, the war was raging in Vietnam and many of the guys back then thought we were all doomed. Country Joe and the Fish summed it up well with their song at Woodstock: "Be the first one on your block, to have your boy come home in a box." Fortunately, I got a college deferment. But one thing I can say about residents of small rural towns is that they join our military at very high rates, fight our wars, grow our crops, and raise our beef. We should be grateful they meet these important needs in our society. There must be a way to reduce the poverty level in rural America to bring about more economic equity. The cost of living is very low, but so are the wages. I am hopeful that the internet, the slower pace in this crazy world, and the ability to work from anywhere will eventually make life less difficult in these areas.

I grew up with country music and have always loved it. My Mom often played Johnny Cash when growing up. I had one of his 8-track tapes in my car in high school. A few of us were also into rock and roll. So, I also had tapes of Led Zeppelin, Rolling Stones, etc. But long-haired rockers smoking pot were generally frowned upon in Edna. Yes, easy listening music like "When a man loves a woman" by Percy Sledge and "Whiter shade of pale" by Procol Harum were popular, because who doesn't like slow dancing? A recent study in the news reported that hugs should last at least 20 seconds to release happy hormones. So, there is some science behind slow dancing. Of course, the Beatles also made it everywhere, including Edna. I remember seeing the Beatles on the Ed Sullivan Show for the first time and the exact location where I saw them. They were reasonably popular until John Lennon said they were more popular than Jesus. A few Beatle records were burned, maybe all 3 or 4 that were purchased in Edna. Long-haired hippies were not well tolerated in Edna. Ironically, 20 years later, long-haired, pot smoking country music fans were plentiful. One thing liberals and conservatives both like is a good party. Maybe, that should be a good starting point.

A fun part of where I grew up was the German-Czech tradition of live music in dance halls on weekends, typically Knights of Columbus halls. BJ Thomas ("Raindrops falling on my head") and the Triumphs were the top area band. He had an amazing voice, of course, and his band mostly played the latest pop hits. These dance halls typically attracted ~500 or so area teenagers and young adults. The younger girls (freshmen, sophomores) often had their mothers watching in the background. The guys typically found out when the girls pushed us away during a slow dance, warning that mom was watching, and sometimes grandma too. The older girls were often outside making out. I suspect there were some heated mother-daughter discussions about this freedom, which the girls typically won by the time they were juniors. Unfortunately, there were always plenty of fights, particularly after the dance when drunks flooded the parking lot. When asked what my biggest accomplishment was growing up in Texas, I often answer "Leaving Texas with all my teeth."

Edna's most notorious badass was Raymond Boyd. He was a small man, about 145 pounds. But he was an unbelievable street fighter. Often bullies from another town would come to a table with Edna folks looking for Raymond, who had quite the reputation. These guys were often a head taller than Raymond. Raymond would pop up, look up into their eyes, and say "That's me! Outside Motherf***er!". He put a few people in the hospital. After graduation, he moved to Houston and quickly won the Golden Gloves Title in his weight class. When I was home from college, I ran into Raymond at the local dairy bar hangout. He had become a professional boxer by then, had grown long hair, was wearing bell bottoms, and hippy beads around his neck. A couple of long-haired surfers from Houston had just pulled up and several high school football players were about to start a fight with them. Being about halfway between Houston and Padre Island (just south of Corpus Christi), this was a frequent "mistake" stop for surfers. Typically, it was two small guys against half a dozen local football players. Of course, the Monday morning story had the opposite numbers. Boyd walked over and said: "What's going on over here. What you got to say about my beads shit-kicker?" The ringleader said, "Nothing Mr. Boyd", and made a hasty retreat along with the rest of his posse. Yes, four or five of them against Raymond might have been a fair fight. But they knew who he was and decided to run. Raymond and the rest of us had a big laugh. Of course, he told us that he would not dare start a street fight these days. As a professional boxer, it could have ruined his career. He developed a national reputation and retired after 12 years with 27 wins (22 by KO) and 9 losses. That sure seems like a high percentage of KOs for a welterweight. Thankfully, Raymond and I got along well. He only fought people who loved to fight as far as I could tell.

Ok, time for a Candy Barr story. Juanita adored my dad and always brought her dry cleaning back home to drop off at his cleaners. When she pulled up in her Cadillac one Saturday morning, Dad immediately told me to run out to the car and bring her clothes in. They were mostly expensive Neiman Marcus outfits stacked to the ceiling in the back seat. I remember following her into the shop on a hot summer day, watching her walk into the cleaners in that tiny Daisy Mae top and tight cut off denim short shorts that left little to the imagination. OMG! The actress Joan Collins was right. She had an amazing figure. Juanita trained Joan in the art of Burlesque for a movie she starred in as a stripper. On this day, I was nearing puberty and remember thinking, "So, that's what sexy women do to men." Good thing I had an armload of clothes covering my midsection. It was an interesting coming of age experience for me. Dad loved her business but sometimes stressed that he might have to pay for one of those expensive outfits if it got damaged. No one in Edna had clothes like her or looked like her without them.

I have no remembrance of my dad's parents, who died when I was very young. My mother's father, Raymond Derrick (Papa), however, had a big influence on me. He never attended college but was one of the smartest and most interesting people I have met. He was truly a "Jack of All Trades". Over his lifetime, he worked in almost every state in the US (47 was the number he once gave) and several countries. Mom told me that he was so valuable during WWII that the locals made sure the army did not take him. He could fix just about anything on a farm. When they built the runway and airplane hangars at the military base in nearby Victoria during WW2, he supervised the work. He ordered all the supplies up front. The story goes that they had only one wheelbarrow of cement left over after completion of the hangers and the runway. He built our house and his own house. I remember my dad saying that the total cost of my boyhood home was less than $6,000. Papa applied for a job in Iran when they were laying another major pipeline in the 1950s. Although he had never done a job of this magnitude, he was hired to oversee maintenance of the entire operation. They simply identified him as the best guy for the job. I remember Papa telling me that several hundred people worked for him. He had a jeep and a small airplane. Oh yes, he knew how to fly too. He once told me he saw a woman in the desert with a bullet hole in her head. It was not a good place for women who disrespected their husbands.

It is very likely that my grandfather was the best marksman in Texas. But he never bragged about it. Dad told me that he came in second in pistol and rifle both at the Texas NRA shooting tournament when he was a grandfather wearing bifocals. He was competing against guys in their 20s with perfect vision. The pistol he used was purchased the week before and he had only shot it once before the tournament. Everyone who hunted with him had a story about how he made an impossible shot. Here is mine. It's permanently etched in my mind as if it happened yesterday.

On a deer hunting trip, we were squatting on the edge of a dense area of forest overlooking a large open field. My grandfather was to my left. A large buck appeared behind us to my right. When we spotted him, he was in a small opening looking directly at us, about 25 yards away. I raised my gun very slowly, but he got spooked and took off at full speed. In the blink of an eye, my grandfather spun around to his left and fired behind us. I remember thinking that there was no way he made that shot. He said he got him, heard him grunt. We followed the blood trail and, sure enough, it was a direct hit to the heart. Through the dense forest, I never even saw the deer as he ran away. Papa later told me that as the deer was running full blast to his right and he was quickly spinning left; he spotted a small opening about 2 feet wide and shot. He said he instinctively spotted the opening a split second before the deer passed through. How could anyone make that shot? I doubt the whole event lasted a second, it took place so quickly.

I was upset when my grandfather divorced my grandmother, Mattie, as I entered high school. She was one of the sweetest women on the planet. But he met and married a woman with similar passions who owned a gun shop/rifle range on the outskirts of Austin. I spent a lot of time there in the summer and became a good marksman. But, after killing my first deer at about age 12, I decided that I simply did not have the stomach to kill animals. It was a shot so far away that my grandfather said my 30:30 was not powerful enough. So, I used his 375 Weatherby Magnum (literally an elephant gun). He told me to aim about a foot above and later stepped it off at ~400 yards away. Luckily, he made sure the kick did not knock me out of the tree stand. I had a big bruise on my shoulder but never noticed until later. The shot went right through the neck and death was instant. I was always very attentive watching him skin and process deer as he pointed out features of anatomy. This may have triggered my early interest in anatomy. His specialty was sausage made from deer and armadillo (pork taste). It was delicious and, thankfully, he always insisted to my mother that it must be thoroughly cooked. I later learned that leprosy is carried by armadillos. Yum! Had it been possum, I probably would have passed. Some animals are just too ugly to eat.

Some of the men from nearby Bergstrom Air Force base frequented the rifle range and witnessed my grandfather shoot. One day they brought a major there and asked if my grandfather would have a shooting match against him. This major was the top marksman at the base and had a standing $1,000 bet with anyone who could outshoot him. That was a lot of money back then. The major asked my grandfather if they could use his pistol. My grandfather said: "Ok, just let me shoot it once to see where it goes". The next 5 shots were dead center. The major did not even shoot. He just took his gun and left. The men said they just won a lot of money betting on my grandfather and asked if he would have been nervous had he known about the bet. He just said "No, it's your money." Always cool, calm and collected.

My grandfather could do just about anything it seemed. He could make a gun from scratch, including hand carved beautiful, raised images of deer or bear on the wooden stocks. He showed me a beautiful metal statue of a deer he made with a welding torch. The anatomical details were amazing right down to their eyes. I also remember seeing a matching pair of new Winchester gold plated centennial rifles on his wall. He said the president of Winchester sent them to him as a birthday present. I often wondered what that relationship entailed, probably some helpful advice. Papa had an amazing gun collection and loved to trade. He stored some interesting pistols in an old, non-functional refrigerator in their kitchen. One was a Civil War pistol that I liked to play with. It was a revolver and needed lead balls. It had a lever that packed balls and gunpowder into the rotating chambers. This was before the time of cartridges. So, there was no way this gun was ever loaded. Or so I thought. One day after Thanksgiving dinner at my grandparents', I retrieved the gun and pulled the trigger while admiring it, as I often did before. But this time it went off and made a hole in the wall, right after my aunt had passed through the hallway on the other side. It scared the hell out of everyone. I had powder burns on the two fingers where the ball passed between. Whew! No missing fingers. Later, my grandfather said he was too curious not to shoot the pistol, so he made some lead balls and loaded it. He needed to know, firsthand, how accurate it was. I sometimes wonder if he was a reincarnation from the Lewis and Clark Expedition. It would have been the perfect experience for someone like him. He taught himself Spanish just from mingling with co-workers. I think he could have picked up Native American dialects quickly too. He never got lost in the woods.

Papa had become friends with LBJ over the years and invited Lynden to his birthday party shortly after he left the presidency. LBJ was feeling ill and could not attend but sent one of his signature hats as a present. It was wrapped in cellophane in a custom hat box. I used to try it on and pose in front of the mirror as a boy. He promised me the hat one day.

I was in college when I heard my grandfather died in an accident. He owned property near Big Bend National Park in West Texas and was dynamiting into a mountain side to build an energy efficient home for retirement. He was going up the mountain on his tractor when it tipped over and fell on him. I learned that my grandfather was not invincible after all. We did not find out about his death until a week later. My step-grandmother, Loretta, said he left nothing behind. We all knew that was a lie. His amazing gun collection and a coin collection that could stock a coin store were supposedly sold before he died. He often showed me bags of Barber silver dollars and half dollars and many gold coins, although they were illegal to own at the time. Most importantly, I never got the hat with the headband label that read "Custom made for Lyndon Baines Johnson, President of the United States". I often think how special that would have been to pass down the family line. Even sadder, Loretta had no children or living relatives. She probably just sold everything.

Years earlier, Papa had survived another accident where he ran a hunting lodge in Idaho. One morning, his Jeep ran off an icy road into a stream below. Someone stopped and managed to pull him from the water and placed him in front of their Jeep parked on the road while he went after the other man accompanying my grandfather. Another vehicle slammed into his car, and it ran over my grandfather. Two weeks later, the headline in the New Meadows, Idaho paper read: "Tough Texan survives!"

My grandmother, Mattie, went to college to get a teaching degree while Papa was away working just about everywhere. Somehow, she sensed that he would eventually leave her. He was "Born a rambling man" as the song goes. Grandma ended up teaching for quite a few years. She had a big heart. After retiring, she fostered two adorable Mexican girls who were living in a violent household. I remember how upset she was when a judge ordered the girls to be returned to their parents. I am not sure how they turned out but the stories of abuse haunted her for the rest of her life.

Grandma was a bit overweight. But rather than losing weight, she decided to fatten up everyone else instead, starting with me. She gave me a big tin of homemade pecan pralines several times a year. And, of course, I finished them off rather quickly, being a kid with no self-control when that tin arrived. Entering junior high, I was a short, fat kid, with glasses. The prettiest girl in the class, Bonnie McKenzie, gave me the nickname "Fat bird". When I run into old classmates, they sometimes call me by that name, with a smile on their face. Fortunately, starting about the time I entered high school, I shot up quickly until reaching 6"1" about the time I graduated from high school. I was tall and skinny. I also had another nickname from middle school, Professor. Apparently, kids in my class knew me better than I did. The problem with me was that I always had way too many interests. I knew that deciding what to be when I grow up would be a big challenge. I loved drafting in high school and to this day think a career in architecture would have been fun, particularly with the green revolution and associated creative innovations. I started college as an engineering major but quickly realized that I could not stomach 3 years of calculus. But for some reason I knew that I was put on the planet to do something special. But what? I kept looking for a sign while exposing myself to as many new things as possible.

I would be remiss if I did not provide a brief description of my immediate family. My sister Delores was a beautiful woman who got a MA in education at Sam Houston in Huntsville and had a wonderful career teaching elementary school. Like our mom, she was sweet and kindhearted. Delores and I attended ballroom dancing classes while in junior high school. The girls wore white gloves, which was practical to counter the sweaty hands of the guys. This was likely the catalyst for my lifelong love of dancing. Delores loved to jitterbug with me and always wanted to dance throughout our lives whenever we visited. By coincidence, her friend Steve (eventually her husband), grew up in Beaumont, where I attended college, and Delores would often accompany him on home visits to see me. He patiently watched as I fixed her up with some of the most handsome guys in my fraternity, including one he knew from high school. Eventually, Delores and Steve started dating, married and had 5 wonderful children. It must have brutal for him to watch the men I fixed her up with, especially those he knew. Their oldest son, Daniel, was ranked the #2 high school quarterback in the greater Houston area his senior year. I often gave him a rough time since "being #2 did not have such a good ring to it." He just said, "You should see the other guy." The other guy turned out to be Vince Young. Yes, the college All American quarterback who single handedly led Texas to the national championship with his legendary 4th quarter heroics and who later played in the NFL (2-time Pro-bowler). Vince turned down the Heisman trophy after it was stripped from Reggie Bush and offered it to him retroactively. Dan played college football for a few years and eventually became an Army Ranger. He is now Lt. Colonel in the Army. He is a sweet man but not someone you would want to confront in hand-to-hand combat, 6'3" and still solid as a rock. Her other children have done very well too.

My brother Chris is 16 years younger. I opened his eyes to the world while he was growing up. He quickly realized that Edna was not the center of the universe and eventually lived all over the world. He attended the University of Texas and is now a professor of economics in Houston. He also teaches critical thinking skills, which sometimes results in death threats from parents who believe he is brain washing their children. Chris was always enthralled with Latin culture and learned much about the history of just about every country south of Texas. To become fluent in Spanish, he taught English in Mexico City for a while. His future wife was one of his students. She came from a prominent Mexican family and attended college in Switzerland. They have two adorable girls who are fluent in English and Spanish and conversational in French and Mandarin because of Chris taking jobs around the world. My youngest sister, Nancy, did not attend college and has had a roller coaster life, particularly since the love of her life passed at an early age. Nancy is much like her mother.

I doubt that either one ever made an enemy in their lives. They never engage in arguments.

I am blessed to have two intelligent and beautiful daughters. Meredith, the oldest, has always been the artistic, creative type. She has been engaged in marketing and modeling. Meredith loved posing for the camera from a young age (**Figure 5**). A funny story regarding modeling involved a photo taken of Meredith wearing a jersey from a Mexican soccer team. The photo of Meredith went viral in Mexico because they thought it was Lana Del Rey and that Lana was a fan of the team (**Figure 5**). She does resemble Lana (also the name of Meredith's beautiful dog). Caroline had always been the easy one to raise, never getting into trouble and doing well in school. She is now a nurse. I love the picture of them in the apple orchard when they were young **(Figure 5)**. I did not marry until age 36 and my wife is 2 years older. We were lucky to have children so late. After dating for 2.5 years, she gave me an ultimatum to either marry her or she was leaving. She was stunningly beautiful and articulate. Thankfully, she did not grow up with Texas grammar, and her mother was an English teacher. She finally got me to lose my bad grammar. This was not so easy. It reminded me of getting better grades than the smart Mexican kids in Spanish class. They could not lose the *Spanglish* grammar that I never learned. **Figure 5** shows 4 photos of my girls.

¿Lana del Rey con la camiseta del León? Presumen jersey alterno de LDA

Resurgió el jersey alterno del León, fabricado por Los de Arriba

Figure 5. The camera has always loved Meredith (~age 3, top left). Girls at apple orchard in their Hanna Anderson outfits (top right). Mexican news photo of Meredith being mistaken for Lana Del Rey (bottom left) and Caroline (bottom right).

Chapter Three
From Nerd to the Guy with the Hot Girlfriend

By the time I reached my senior year in high school, I looked very different and pretty girls started to notice. I had begun to grow my hair a bit longer and often wore bell bottom pants. My friend Chuck and I were infatuated with the excitement emanating from Houston, 100 miles away. We loved rock music. We would sometimes spend the weekend with Chuck's cousins in Houston. One of his cousins was cute and we barely avoided getting caught by her parents making out. It was an amazing scene being in Houston at that time. Alans Landing, downtown, was filled with hippies, hippie "wanta-be's", and great music. Smoke was everywhere, sometimes tobacco and other times not. We saw the movie Woodstock my senior year while in Houston. It was a bit dangerous to grow long hair in Edna. On my trips home from college, I got quite a few stares as my hair got longer. On one occasion, I decided to get my hair cleaned up a bit but insisted to my boyhood barber to leave it longer. He turned me away from the mirror and started cutting. When he turned me around to look into the mirror, I literally had a burr cut. I was mad, but it grew back. Rock was making a bigger impact in nearby Victoria, which had about 50,000 people at that time. Chuck and I started following a new band named ZZ Top. They sometimes played in Victoria and we helped them load their van after a concert once. Somehow, I knew this band was the real deal. I believe we paid about $1.50 to see them in those days. It's hard to believe this band is still performing over 50 years later. But they charge more now.

By the end of my junior year, I started dating attractive girls from out of town. A couple were cheerleaders and among the most popular in their towns. But for me, the girls in Edna still saw me as "Fat Bird." There was a lot of bullying going on when growing up in Edna. It was considered "sissy" for a guy to make good grades. It had been so many years since a boy graduated Valedictorian that I just got a plaque inscribed "High Ranking Boy." It was ironic that some of those bullies later begged me to fix them up with pretty, out-of-town girls. I dragged my feet on those requests. I was reasonably popular by the time I graduated. Maybe it was not so good to be among the *most* popular since there were a few suicides involving those individuals from my class in later years.

Growing up, I was under a lot of pressure to play football. Football was king in Texas. I was an average athlete at that time and was smaller than other athletic classmates until after I graduated. Being among the youngest in my class and entering puberty later did not help. Of course, my older cousin, Forest Wiegand, was a high school "All American" and took Edna to the State football championship. He later was the starting center for the University of Texas. They only had 2 losses during his 3-year tenure as starting center. One was his first game and the other was his last, a loss to Penn State for the NCAA National Championship. But in between, he got a national championship ring. When we ran into him in later years, he always let us try on the ring. Forrest never played in the NFL because he was too small at 204 pounds. But he ended up coaching some successful high school teams. I often noticed the aches and pain from his football days that followed him the rest of his life. My cousins Daniel and David Hollas were also starting quarterbacks for Edna. Daniel was pretty good despite being a polio survivor with a shorter leg as a result. Both are among my relatives who became college professors. So, as you can see, there was considerable family pressure on me to play football.

In the summer before starting my freshman year of high school, I got the mumps and was bedridden for nearly a month. About 2 weeks after summer football training started, my doctor proclaimed that I was well. I could go play football! It was August in Texas, and I was so weak I could barely walk. The others were already in good shape by then. I remember that the coach had them run blood drills at me shortly after my arrival. For those not familiar with football, with this drill you try to go forward while other players come at you one after another. The best players make forward progress. All I remember is the ringing in my head, the pounding my body took, and how much bigger these guys were. Thinking back, I could have died of heat exhaustion alone. I endured a few more days before telling my dad that I was quitting football. My Dad was very disappointed. In later years at class reunions, I remember how less intimidating these guys were as I had become their size or bigger. My Dad did not really know how to handle my academic success. It was totally unexpected. Later when it was announced in the local newspaper that I was valedictorian, many people came by the cleaners to congratulate him. He was very proud and finally told me that I was right in choosing a different path. He apologized for all the pressure he put on me. Ironically, Dad was not a very good athlete. After school, he had to go home and work on the farm. None of the 11 kids in his family ever had the opportunity to play organized sports. So, he was really trying to live the life he was denied through me. I got it. And, because of my decision, "I ain't got no dain bramage."

My best friend at St. Agnes elementary school was Dale Martin. His father was voted "Most likely to succeed" at the University of Texas Law School. Dale senior had once won a million-dollar lawsuit but was penniless when he moved his family back to Edna to live with his mother. He and his wife, Isabel, were alcoholics. It was disappointing to sometimes see his dad ask Dale Jr for money he made working at a gas station. They lived across from St. Agnes church and school. We sometimes watched TV game shows when at their house. Dale senior could literally answer every question before any of the contestants. Dale junior was a real schmoozer and would often visit the nuns across the street on weekends. I sometimes accompanied him. I remember him saying, "Sister so and so, why didn't you get married. You are so beautiful." Dale always got slightly better grades than me, A+s to my As. The nuns did not keep grade books. They just gave subjective grades. But I knew I was outperforming him. When I told my parents, they did not believe me for the longest time. It seemed obvious that Dale's brilliant father would have the superior son. But when we got a lay teacher in the 6th grade, she corrected the situation. I must admit that it was disappointing that my parents did not believe me. I understood their thinking though. The lessons for me were that life may not always be fair and that you must prove your worth and be persistent in that pursuit. Years later, Isabel would often brag that Dale Sr died of a heart attack with a smile on his face while making love. They were something else and a very handsome couple.

Edna was filled with some interesting characters. There was another guy a few years older nick-named "Tarzan". He had long blond hair, lived alone in the country, and obviously liked his privacy. He was a handsome guy who looked like Ron Ely who was the 1960s version of Tarzan in the movies. But I never saw him with a girlfriend. I also don't seem to remember anyone named Jane while growing up. Where is Jane when you need her? Apparently, he was well off and did not work. No one really knew what he did with his time. I believe Chuck and I may have been his only friends. We enjoyed listening to his perspective on life when he came to town. He really loved nature.

I would later get another lesson in life's unfairness. My senior year I built an electronic computer for the science fair. This was before Texas Instruments came out with their pocket calculators with the red numbers. There was no such thing as a home computer at the time. There were large computers at colleges where you learned coding and fed a stack of cards into the computer. I recently learned that Bill Gates and his high school friends were writing software for large main frame computers for developing tech companies at this time. The experience eventually led to their goal of developing personal computers for working people around the world. This became a reality shortly after I graduated from college in 1974. Boy, would it have been amazing to be part of Bill Gates' high school group, which included Paul Allen. But I was on my own in Edna.

I got the idea for building a computer from an article in Scientific American, where they explained "and, or, not" (Boolean) logic. The article also provided a schematic for basic addition, subtraction, and conversion from base 10 to binary. I visited a professor at Texas A&M who gave me some advice and specifics on newly developed integrated circuits and transistors that I could use for the project. I mounted everything on a large, perforated board and worked on the project for many months. I won 1st place in the physics division at the local science fair. But I lost the grand prize to another kid whose brother, a petroleum engineer, did the project for him. He did fractional distillation of crude oil, collecting different by-products at their specific boiling points and placed them in labeled vials. The judge was a petroleum engineer and was clearly impressed with his project. Ironically, he thought that I could not possibly have built the computer on my own. Mr. Cranford comforted me in my disappointment. But I got the last laugh. My project made it to state finals, while the other guy never made it past regional.

I also engaged in slide rule competition and was reasonably competitive. For those who don't know, a slide rule is an analog computer consisting of three parts (usually wooden or metal) labeled in a logarithmic manner with the outer parts fixed and the middle part sliding between them. There was a girl who always won State. Slide rule was the center of her universe. At State competition, she once sat in front of me to my right. Near the end, I just put my slide rule down and watched her, knowing I had no chance of winning. She had a slide rule in each hand sometimes and was unbelievably quick. It was amazing to watch. A few years later, slide rules became obsolete. I sometimes wonder if she burned her slide-rules, committed suicide, or just landed a job for the future like a telephone switchboard operator. On the drive to State competition, the football coach took me and Bob Shaw to Austin. Bob was one of the fastest guys in the state and was competing in the 100 yard-dash. We did not have much in common, but he was a nice guy and I gained some knowledge about how top athletes train. We were the only two who made it to State level from Edna that year.

Texas also had Number Sense competition. Sometimes the math problems covered nearly half a page, and everything had to be done in your head. You could only write down the answer. I tried that briefly but never got into it. But, at district competition, our guy was sick and did not show up. So, the teacher asked me to give it a try. Scores could be low because you got -10 for the wrong answer and only +5 for the correct answer. It was a very difficult test. I got twice as many correct as incorrect, but that was a 0. I placed third out of about 25 students competing. I could have just handed in a blank test and done that! But the math coach was pleasantly surprised. I am grateful they did not announce our scores when reporting the awards.

Growing up, I enjoyed watching baseball games with my dad and his friends on Saturday afternoons. It was so exciting when Houston got a franchise, the Colt 45s, when I was 10. My Dad took me to a couple of games a year. I saw their first no-hitter, by Don Nottebart. They were so bad at the time, they almost lost that game too. One of my dad's friends, Basil Green, often dropped by our house for Saturday afternoon baseball games. He owned the Edna Restaurant. Basil fashioned himself as the classiest man in Edna, smoked from a cigarette holder, always wore a pressed white shirt with the sleeves rolled up, and nice dress pants. He was one of my dad's best customers. But he was a bit of a racist and did not like betting on any team with black players. So, my dad got teams with players like Willie Mays, Bob Gibson, and Hank Aaron. Dad usually won those wagers. Afterwards, my dad sometimes laughed with me about how stupid Basil was to not accept some of the best players in baseball, simply because of their skin color.

I remember watching the Astrodome go up near Colt 45 stadium. We all thought it was going to collapse from all that weight over an open space. But the architects knew what they were doing. My Dad went to the first game where New York Yankee Mickey Mantle hit the first indoor home run in a major league baseball game (preseason, but still a major league game). He took me to the second game. I remember walking from 90°+ heat into 72° air conditioning and seeing this huge expansive indoor space. It was a surrealistic, magical feeling and among the most lasting memories in my life to see "The Eight Wonder of the Modern World". It is notable that some of the biggest changes in baseball occurred because it was just too damn hot to play baseball in Houston. It was likely for this reason that Southern franchises were late to be founded. Although MLB had experimented with night baseball prior to this, it finally caught on out of necessity for the Houston team. Prior to the Astrodome, I remember going to a double-header in Colt 45 stadium in August. I later heard that the Houston catcher, who caught both games, lost 20 pounds. Of course, the unbearable heat and humidity was probably a home field advantage when Northern teams came to town. A problem with the Astrodome was that grass would not grow well inside. Developed by Monsanto, sales of their artificial grass took off and the product was renamed Astroturf due to the success created in Houston. Houston gave the world night baseball, indoor baseball, and fake grass. Initially, they did not consider that having a white roof might interfere with tracking the ball on defense. So, they later painted the roof a darker color for contrast. This contributed to poor growth of the grass, which led to installation of Astroturf. Here is another interesting bit of trivia regarding the Astrodome. On June 15, 1976, a game in the Astrodome, an indoor stadium, was rained out! People could not get to the stadium with over 7" of rain in a short time. The Astrodome field was also 45' below ground, resulting in flooding of the field, lower ramps, and entries.

After I got my driver's license at age 14, I became the designated driver while my dad and a couple of fiends got "warmed up" with a few beers on the drive to games. Texas had no open container law at that time. I remember them once raising their beers in a toast to a policeman who waved back as we pulled out of a gas station while passing through Rosenberg.

Upon graduation, I could not wait to move to Houston and spend the summer there before I started college. I contacted some guys I knew who graduated just before me and we all decided to rent an apartment on Richmond Ave just across the street from one of the hottest night clubs in Houston. This was a huge apartment complex with multiple swimming pools where many young people would party like crazy all weekend. It was a big change from Edna and an absolute blast. I turned 18 that summer.

Immediately after my arrival in Houston, I landed a job working in the darkroom of a photographer who did commercial work. Prior to that he was a professional wrestler. I heard some crazy stories from him. He told me how he would sharpen one fingernail, make a tiny cut on the opponent's forehead and smear the blood all over his face to make it look much worse. It was all planned with the other wrestler's consent. He also hooked me up with a leading wedding photographer in River Oaks, the most affluent area of Houston. After tagging along with their other photographers and taking pictures alongside them, the studio owner was soon sufficiently impressed with my ability to let me do weddings on my own. I must admit that some of his wealthy clients appeared to be surprised about my age, but I knew the drill by then and just took charge. I never heard any complaints, and he kept assigning me to more weddings. A perk in this line of work was meeting pretty, rich girls at weddings. I was invited to a few upscale parties after some weddings. It was a whole new world for me. I had no concept of wealth. Some days, I had 3 weddings in a row with little time in between. I was wearing a suit and had a mostly nonfunctional AC in my car. I must have looked like hell by the time I was done. But it sure beat the other summer jobs I had later. He only paid me $25/wedding. Annoying, and I knew it, but 3 weddings would pretty much cover my monthly share of the rent. Heck, I was living in the heart of Houston when it was the hottest spot in America.

In the summer of 1970, young people were flocking to Houston for jobs in droves. I remember hearing that 1,000 new cars were on the road every week in Houston, growth largely sparked by the oil and gas boom of the 1970s. It was in the middle of the Vietnam war and a very crazy but exciting time to be there. Back then, the night clubs were lax on IDs. I almost always got in, and I did not even have a fake ID. Note, shortly after I turned 21, they changed the drinking age to 18. The logic was that if you could get your ass blown off in Vietnam, you should be old enough to buy a beer. After the war, the drinking age reverted to 21. The timing never worked to my advantage anyway. This is a good time to discuss Constitutional Amendments. Did you know that the 18th Amendment outlawed alcohol and the 21st legalized it again? Easy to remember since it's illegal to drink at 18 but legal at 21.

I often flirted with a beautiful 21-year-old by the pool on weekends. She was the first Miss "So and So" that I dated and grew up in a town that is now a suburb of Houston. She always gave me a rough time about my age, but it was clear she liked me. After a few weeks of verbal banter, she invited me to her apartment one Saturday night after the pool party. I left late that night and gave her one of my wallet-sized, high school senior pictures. The next day when I went back to see her, she had framed it and put it on top of her TV. But, when she learned that I was moving away to go to college at the end of the summer, she lost interest. I have included that photo and a few others later in my life in **Figure 6**.

I met a few other interesting women that summer. One worked at the left-wing underground radio station, Pacifica. On May 12, 1970, shortly before I moved to Houston, it was bombed by the KKK because of their liberal viewpoints and bombed again that Fall. Most of the guys there dressed and looked like Fidel Castro (beards, green fatigues and his signature hat). In between those bombings, I visited her at the station once that summer. After learning about the previous bombing, I saw no reason to press my luck by returning. She also worked with a company that sponsored local concerts. So, I got to go backstage a few times. The best summer concert was the "Day of Joy" at the Almeda Speedway. ZZ Top was just starting to get national recognition. They were there with several other nationally known groups, like Albert King, Leon Russell, Rare Earth along with several bands with current Billboard hits (mostly one hit wonders). I have included a picture I took of ZZ Top that day **(Figure 7).** There was a truce between local police and kids at the concert. State Troopers watched in disgust as the air filled with marijuana smoke. I remember lying back near the front row with my eyes closed enjoying the music on this hot summer day. I was awakened by an object hitting my chest. I had just been run over by a watermelon being thrown out by the organizers! Other than being run over by a watermelon, it was a peaceful, wonderful time filled with great music.

While in college, Willie Nelson had some wonderful July 4th picnics, often on a farm that was a reasonable drive from my parents' house. I took my mom to one and we had a blast. Dad had to work late and could not make it. Of course, there were many joints being passed around. I caught my mom indulging. She was a free spirit at heart. Mom claimed she didn't feel anything, but she certainly was laughing a lot. That may have been one of the most exciting days of her life. Don't forget to spend time with your parents when you are young and single. As she was dying of pancreatic cancer years later, she tried it again and said it helped make her happy and numbed the pain. I lost both parents a few years ago. I sometimes must stop myself from calling them, realizing it is no longer possible. It is difficult losing your grandparents, but you also know that is part of getting old. Losing your parents is different and it comes with the recognition that you are next in line. I think about them every day and it makes me smile. Memories are so much more important than things.

Figure 6: Senior high school picture. Scuba diving days in FL. Recent work picture.

Figure 7. Photo I took of ZZ top at the Day of Joy festival in 1970. Left to right, Dusty Hill, Billy Gibbons and Frank Beard, who only briefly had a beard.

Chapter Four
College Years

There was not very much good advice in Edna for someone like me regarding where to attend college. The school councilor pretty much directed most students to junior college while others stayed on the family farm, became beauticians, auto mechanics, carpenters, plumbers, insurance salesmen, opened restaurants or small businesses, etc. The high school troublemakers often became policemen. Most students interested in college started at Wharton County Junior College or Victoria Junior College and later transferred to a regional four-year college if they made it that far. I was leaning toward attending the University of Texas. But I was discouraged by many. "It's too big. You will just be a number." I interviewed a few places, including Rice. While there on interview, I remember seeing a guy crash into a door while walking and reading a book. I left the Rice campus believing that I was not nerdy enough for that place. They made the decision easier by not extending an offer. In the end, a friend of my dad's talked me into going to Lamar University. It was a manageable size, with about 10,000 students, and had a nationally ranked engineering school, which was my initial chosen major. However, I found calculus boring and the thought of taking more did not excite me. I always loved biology and switched majors by the second year. My family doctor had encouraged me to become a physician since I was a kid. I never really saw myself in that role to be honest but took pre-med courses just in case. Becoming a scientist just seemed like more fun even though I had never done any organized research.

I initially lived in a college dorm with a dorky roommate named John. I became good friends with Bob, the guy across the hall who had a baseball scholarship. He was a tall, handsome guy, a real girl magnet. Bob liked to play jokes on my gullible roommate. Once John brought a fifth of vodka to the room. While he was gone, we poured about 2/3 into another container for us and topped the bottle off with water. John got very drunk that night and bragged the next day that he "put it down like it was water" while laughing along with us. Bob and I decided to join the same fraternity our freshman year. It was a lot of fun with all the parties and sorority girls. An additional benefit was living at the frat house for $35/month. So, I moved there after a freshman stint in a dorm. I managed to support myself largely on shooting weddings, low rent, low tuition, and hard work during the summer. Unfortunately, living at the frat house was a bit distracting. For instance, I was kidnapped with 2 other actives by pledges the night before a big comparative anatomy test. After being blindfolded, we were dropped off on a gravel country road on the outskirts of town. I just purchased a new pair of boots that were not broken in yet. We got back about 3 AM after hitch-hiking home. My feet were covered with blisters. I had put off studying until the last minute. I would not recommend this lifestyle for anyone who is pre-med. Fortunately, I managed to end up with about a 3.5 GPA, good enough for grad school and with stronger science grades. I sometimes wonder if I could have had a 4.0 if not for girls and fraternity distractions. I remember a physics class with about 100 students and making one of only two A's given in the class, which was almost all pre-med majors. Physics has always been one of my favorite sciences, thanks to my high school teacher Evies Cranford. I enjoyed learning that physics provided mathematical formulas explaining many of life's physical phenomena. I withdrew from Comparative Anatomy of the Vertebrates on the first pass because I had a 75 average. I later found out that would have been one of the few A's in the class after a curve that prevented a likely 80% flunk rate. The teacher earned the nickname the "Big Bad Wolf" because he was so ridiculously hard. It was well-earned. You could never study enough, so many animal phyla, so little time. Comparative anatomy and organic chemistry were the weeding out classes for premed students, who always seemed to be the main competition for me. On the first day of organic chemistry class, the professor told everyone that only 25% of the class would make it through both semesters with a passing grade. A classmate joked that the teacher was at her best when she lectured drunk. I agreed that she was certainly entertaining while wondering when she was going to trip and fall on her way to the blackboard. There was no grade padding back then and it was very difficult to challenge a professor with a poorly written question. I made it through both semesters of organic.

But it was the class that soured me on chemistry. I felt like I had seen enough carbon bonds for a lifetime.

Our fraternity was a surprising mix. We frequently won the highest frat GPA award with physicians, lawyers, dentists, engineers, and a scientist (me) coming from our group. We also won the beer drinking contest every year. In the beer drinking contest, two guys from each fraternity would compete. They drank beer until only one team was able to continue drinking. Throwing up was a disqualification. You could not leave and go to the bathroom. You had to pee in your pants. It was interesting seeing that look on their faces when contestants finally relaxed and let it flow. Then, the discomfort gradually set in as their pants cooled off. It usually took about a case of beer to win and there was a time limit.

At a fraternity meeting, our president suggested that we should sponsor a golf tournament of celebrities and NFL players and declare Beaumont the "Pro Football Capital of World". The initial response was: "What! We don't even have a NFL team?" At the time, Beaumont had 16 NFL players, which was incredible for a town of only 100,000 people. Sports Illustrated bought it and did an article on the event. Apparently, the idea was resurrected later with the establishment of a Professional Football Capital of the World Museum Foundation in Beaumont. Our frat also sponsored a non-stop softball tournament fundraiser with kegs of beer provided. I remember standing out in right field, drunk and tired, and watching a ball sail over my head at about 4 AM. Luckily, our frat attracted quite a few members of the college baseball team. So, we usually won the tournament. I am a pretty good contact hitter but a terrible fielder because of a weak left eye and poor depth perception. Coaches always put me in the outfield where this visual impairment was magnified, which I hated.

While growing up, I never got to play Little League at age 12 because of my birthday being just days before the cut-off date. My last year of Little League I was getting better but played almost the entire year with an infected foot. Our family had been picnicking on the Navidad River near my grandparents' house. There was a wooden bridge, and I was running around barefooted. I got a splinter in my foot. Mom took me to the doctor for removal. Apparently, he left some of the splinter behind and my foot was draining pus much of the summer before I finally went back to have the rest removed. So, I was running around playing Little League baseball most of the summer with an infected foot. I had a similar memory a few years later while visiting my grandfather with an ingrown toenail infection. His male dog fell in love with my foot and showed considerable affection for it. Dogs usually like me but he got a bit carried away.

I tried out for Senior League Baseball the next year at age 12. When I stepped up to the plate, they put in a guy named Armando to pitch for my tryout at bat. Although he was a year younger and also trying out, he showed me this nasty curve and an unhittable fastball. I had no idea how to hit that stuff. So, I quit baseball too. Armando became one of the best high school pitchers in the area by the time he graduated. Bad luck of the draw for me that day. I ran into him at a bar while visiting home years later. He regretted passing on several college baseball scholarships. He was a bit insecure scholastically. But he said his daughter attended college on a softball scholarship and did well. Apparently, genetics can be helpful in sports, for other people anyway.

Back to my college days. Through my friendship with Richard King, my little brother in the fraternity, I got to know Louie Kelcher well. Louie would go on to become an All-Pro defensive tackle for the San Diego Chargers. Yes, another NFL player from Beaumont. Richard and Louie went to high school together. Louie became a football star at SMU. We often hung out at the beach during the summer. Louie was 6'5", weighed 285, wore size 17EEE shoes, and had gigantic hands. I remember once asking Louie if he needed another beer. He turned his hand around and showed me a beer that was buried in the palm of his huge hand and said "no thanks, I have one." Before being drafted and getting his first big paycheck, Louie married his high school sweetheart and asked me to take the wedding photos. Louie was one of the original "big men" in the NFL before the time of steroids. He got the largest tuxedo available in Houston. Nonetheless, it was so tight that the pants zipper broke. As one of his fellow team members was on his knees trying to pin the zipper shut, I got a picture at Louie's back. Without that background information, a different story line comes to mind. When I showed the picture to Louie, he said: "If anyone sees this, you're a dead man." Of course, I destroyed the negative and print. Sadly, it would have been a source of amusement in later years as Louie became a big celebrity in San Diego and would have likely found it amusing by now.

There was a nearby town, Vidor, famous for racism. No blacks dared to live there. We would often see their KKK chapter burning the cross at the close of Friday night local TV news. In later years, 60 Minutes did a report on a black family brave enough to live there. There was still a lot of racial tension in those days, largely sparked by forced integration of the schools. During my time at Lamar, I never once dated a woman from Vidor. Nothing in common. But I did go out with another "Miss so and so".

One summer while home from college I worked on a road construction site in Victoria for a company run by my dad's best friend. I was amazed at the work ethic of the mostly Mexican workers, often over 50, who had been doing this hard work for decades. I doubt many lived to retirement age. I just wanted to make it through the summer. I became good friends with a Mexican man in his 50s who had around 8 kids. We both loved baseball. He asked me to come over to his "shack", have a few beers, and watch a Houston Astros game on his new color TV. He was so proud that an educated white kid would do that. I had so much fun. His kids were adorable. I suspect that most graduated from college since that was a clear goal of their dad. We need these wonderful people. Immigrants built our country. I wish more people could walk in their shoes and see what their life is like. I did, and it is not an easy life. Many may not be fully aware of their contributions. Did you know that the most highly educated immigrants to the US are black? Yes, Nigerian immigrants have the highest percentage of college graduates. Did you know that Muslims are only 1% of the US population but between 5 and 10% of the physician workforce (difficult to find precise data)? So, Muslims are more likely to be your doctor than a terrorist. Immigrants account for over 23% of all inventors listed on US patents. They pay far more taxes than they receive. The US debt would be much greater if not for immigrants. I could go on but will stop at that.

While at that construction site, a pretty girl kept driving by and flirting with me. One day, she asked if I wanted to join her after work. Of course, I was dirty when she picked me up. She drove to a secluded lake on the edge of town, quickly took her clothes off and jumped into the water. I got the hint, needed a bath, and joined her. About the time things got interesting, a family pulled up with their young children. We politely walked back to the car, naked, while watching the parents trying to cover the kid's eyes, and then quickly drove away. She eventually drove me back home. I never saw her again, but it was a memorable day for me.

Aside from all the distractions, the nagging question of "What am I going to do with my life" was always near the surface. I learned in college that my calling was somewhere in the field of biology. Ultimately, I decided to pursue a PhD in Anatomy at the University of Texas Medical Branch at Galveston (UTMB) because I am visually oriented (photography), enjoyed anatomy, and loved histology class (looking at tissue slices under a microscope). I was very good at mentally imaging cellular relationships from 2 dimensions (histology slides) to 3 dimensions. It seemed like the best direction for me. I knew I would enjoy teaching anatomy and was excited about exploring research. I had a $6,000/yr stipend and free tuition. I probably could have lived on that in those days if I gave up dating. But that was not an option. I also loved to go dancing. Those were different times. I remember that $20 was enough to pay for 2 or 3 large bags of groceries, including meat. Mixed drinks at high end bars were less than $1 and there were always beer specials during the week, 25-50¢ a beer. You could find a shared rental for about $100-200/month each. I did get audited by the IRS regarding my stipend. Apparently, I had one that was not deductible like the other grad students in my class due to one word in the title of the stipend. I had to pay ~$200 more in taxes. Things have not changed much, have they? It's so easy to go after the little guy.

I once went out to the most popular night club in Galveston when they had a country dance contest. Galveston was pretty much a surfer town, so I had a chance. On that night, I had a date with the prettiest woman at the grad school. We won! But I must admit that the audience was probably so busy staring at her they probably never noticed me. She was from a wealthy family, and I was paying for all the expenses, but she insisted we split the prize money. Oh well, it was worth a try. I also got the DJ to play my favorite country dance song for the finals. That was before rigged elections became so popular. Ahead of my time!

I also continued to shoot a few weddings back home, about 1.5 hours away. By that time, my dad had started a carpet cleaning business. He was nervous about flying to Denver to get fully trained by the equipment company. So, he sent me and I helped him set up the new business one summer. A while later, he gave me a small portable carpet cleaning unit to make some money on the side. This was long before stores began renting these units to customers. I managed to land a big job at a local hotel on Galveston Beach cleaning about 200 rooms during the off season. Quite a few times I would work all night, go home, shower, and attend classes after not sleeping. It took a couple of months to finish the job. I was likely the only grad student to work multiple jobs while attending school. It's easy to see why that is frowned upon and generally not allowed these days. Thankfully, I was able to easily maintain the required B average for graduate students despite all the diversions.

I often got some nuisance phone calls while living in Galveston. "How much is yo pork chops today?" Or "When do the peaches get here?" In the phone book, my name was listed just before Gerland's Supermarket. My number also differed by one digit from a popular black bar. Sometimes, I would get a call at ~2am in the middle of the week from an angry black woman. "Is Leroy there? I know he's there. Tell him to get his black ass home." In retrospect, it was just a reminder that we are all part of the same story, just passing through life, often not paying enough attention to small details.

On the way back to Galveston one day, I got stopped by a policeman in Kendleton on Highway 59. Kendleton was so small that it would probably qualify as a ½ horse town. He took me to the police station and they demanded I pay a cash fine before leaving. This speed trap was so notorious that it was later featured on "60 Minutes" and shut down as a result. Drivers were typically speeding at about 70 mph when all of a sudden the speed limit dropped to 35 for no apparent reason. Just about every other person I spoke to who frequented this route had a similar story. There is no telling how much cash these guys pocketed.

During the 1970s, Galveston beach was probably at its all-time peak with so many young people flooding into nearby Houston for jobs. Well, except before the 1900 hurricane, which is still the largest natural disaster in US history with up to 10,000 deaths by some estimates. After that hurricane, they initially weighed down many of the bodies and sank them offshore. But many washed ashore. There were so many bodies on the beach that hundreds of funeral pyres burned for weeks on the beach to cremate them all. Before that storm, Galveston was among Americas richest cities, based on per capita income. There are still some nice mansions that survived all the hurricanes. A more recent hurricane wiped out many of the old trees, which was sad, but it made the Victorian houses more visible. Apparently, excessive standing salt water from flooding killed them. Unfortunately, Galveston, along with many other historical cities like New Orleans and Manhattan may not survive this century due to climate change.

When I lived there, one of my favorite places was a bar on the beach with great live music on weekends. I was often among the first to get people dancing in the sand. The house band was named "Fat Cat" and they played mostly Southern rock. I would sometimes meet a girl in a group of 3 or 4 from out of town and the group sometimes ended up at my place (an apartment in an old Victorian) on Sunday evening where a few of my med student friends joined us. In effect, I saved them study time by being their wingman. Donnie, who later became a surgeon in Branson MO, serenaded us with his guitar and singing. He had hair down to his waist before having to cut it when moving on to the clinical years. My roommate, Jerry, later became a brain surgeon. He ended up marrying Sandy, one of the women who hung out with our group. Those were the days! Yes, burning the candle at both ends. But I would argue that I have always had a well-balanced lifestyle. To this day, I believe that enjoying life to its fullest and treasuring every minute on this beautiful planet has kept me sane. And dancing is such great exercise.

I still love to go dancing at age 74 and am often the one who gets people dancing. Some things never change! Typically, I ask several "older" women (usually in their 50s) who were dancing in place, to join me. Later, younger women my daughters age sometimes ask me to dance after watching. It is sad that so many men just stand on the sideline while many women clearly just want to get on the floor and dance. Music and dancing are a big part of my life. I probably have a better play list on my cell phone than many of my medical students. I feel blessed that young people have always enjoyed my company. My daughter's high school friends enjoyed having conversations with me when they came over. Since so many of them went on to med school, law school, or grad school, I suspect they enjoyed the deeper conversations.

The US has a negative stereotype of aging. Many studies show that staying young at heart increases longevity. I would recommend the book "Breaking the Age Code" by Becca Levy. Age negativity is a major contributor to shorter life expectancy in the US, compared to countries like Japan where aging is a positive experience. Although I quit running about 20 years ago, I walk between 5 and 10 miles a day. I may be the first to walk to their grave at age 110 and jump in while giving it my last dance move. Research shows that couch potatoes die sooner than more active people. My advice: Get your ass off the sofa and go for a walk or turn on some music and dance! "Dancing with myself, yes I'm dancing with myself… (Billy Idol)", when necessary.

My roommate Jerry was a child genius with an IQ of about 220. He could literally read a medical text in a few days and answer anything written in the book. He was a speed reader with eidetic memory. I really enjoyed playing chess with Jerry. But we became so evenly matched that we finally grew bored and gave up. He started med school early, 19, I believe. Sometimes it appeared that med school was a bit too easy for him, even boring. As is often the case with super smart people like him pushed into classes with older students, he was a bit socially awkward. One day he returned from a walk on the beach and announced he was dropping out of med school and moving to Kansas City. Apparently, he met the "girl of his dreams". I believe she was a waitress, about 20. When I asked him, "Why not have her move here? There are plenty of similar jobs here." He responded like that was a revelation, "Great idea!" I believe she left the next day and he remained in school and never saw her again. He later settled on Sandy from our group and they eventually got married. I remember going out with Sandy before Christmas (as a friend). When Jerry came back from break, he asked me not to date her anymore because he was in love with her. Of course, I was so happy for them. They later married as Jerry finished a neurosurgery residency and we have been lifelong friends. The added perk was that Sandy seemed to attract very pretty and intelligent girlfriends who liked to flirt with me.

As I have gotten older, I have come to realize that excessive beauty can be a curse for both women and men. Members of the opposite sex may literally worship the ground that beautiful people walk on. As a result, they often expect preferential treatment and become obsessed with themselves. Even worse, they may treat others who are less attractive disrespectfully. I remember an incident at the hottest night club in Tampa where I overheard a conversation in the restroom between two very handsome young men in their 20s. They made a $20 dollar bet to be the first one who got laid by an ugly woman. Panties would be proof. A short time later, I remember seeing both men leading a qualified woman out the door. How can human beings be this disrespectful of other human beings? Fortunately, I have seen beautiful men and women pass through med school and become some of the most compassionate, kind-hearted doctors you would ever want to meet. The world is filled with a wide spectrum of people. I guess if you didn't go out with a few assholes in your life, you may not fully appreciate the special ones. I am not a perfect person by any means but have always tried to treat everyone with respect and kindness. If I don't make a few people laugh or smile every day, it would be a bad day for me. I sometimes try to make people laugh who look sad and distressed as I walk by, often an elderly woman or man. "Wow, you nailed it! Keep it up. Good job!" They give me a puzzled look. Most often, they smile back, probably the first time that day. They usually realize I was messing with them but appreciated that I cared enough to try and make them smile.

Before graduation from UTMB, Sandy joined me on a trip to Mexico with two women I knew from Minneapolis who flew down for Christmas break. They insisted on going to Mexico because it was still too cold in Galveston. I had no idea what it must feel like to escape -30° weather at that time of the year having spent my entire life on the Texas Gulf Coast. So, I drove these three women to Brownsville, and we walked across the Mexican border to Matamoros. The first stop was an outdoor tequila stand where we promptly sampled about 6 or more shots apiece, trying different brands. I remember it was a bright sunny day and we were quickly buzzed. One of the women noticed a pottery place across the street and we went browsing for a while. After they accidently broke a vase or two in their drunken stupor, the owner demanded we purchase something. One of the women from MN picked a large vase about 3' tall that I ended up carrying around the rest of the day. We ended up in a bar filled mostly with men and a few women. Suddenly Sandy got up and knocked the shit out of a tiny Mexican woman sitting nearby. Sandy, fluent in Spanish, apparently got mad at the woman talking about "Those American whores." I realized later that the only women in this bar were prostitutes. That was the *first* bar we got kicked out of. Luckily, as the evening progressed, I managed to keep everyone out of jail, barely. We were still pretty high when walking across the border that night. As I remember, one of the MN women was on my shoulders with the vase on top of her head. We probably resembled a wobbly, walking totem pole. I wish I had a picture of that! A Mexican border guard began asking her questions, "What do you have in there, drugs?" She cussed him out and he responded, "Don't mess with me or I will throw you under the jail!" I apologized for her drunkenness and managed to get everyone across the border and back to the hotel room safely. I quickly found out that the woman who insulted the border guard had some pot and pills in her purse. Things could have turned out so much worse. I did not invite them back for another visit to Galveston. I was surrounded by more than my share of crazy people already.

I became good friends with a couple of local guys who lived next door, Jim and Chino (nickname). Chino moved in with his girlfriend about the time that Jerry left for summer break. So, Jim moved in with me temporarily. Upon return, Jerry decided to move in with some other classmates. One night, Jim and I parked our car and headed for a local bar. We were immediately jumped from behind by two guys. I remember being kicked in the ribs by one after being knocked to the ground. They left as quickly as they arrived. I asked Jim if we should call the police and report the assault. He told me, "They were the police." It turned out that these off-duty policemen attended high school with Jim. They never got along and Jim flipped the bird while I was busy parking the car.

There was another episode in Matamoros with my roommate Jim. His girlfriend had just broken up with him the previous night and he was very distraught. By early-afternoon the next day (Saturday), he was drunk and stoned when he decided to drive to a whore house in Matamoros. I could not stop him. He was a big strong guy, 6'3" and about 220 pounds. So, to make sure he did not kill himself on the road, I drove him to Mexico. Shortly after our arrival, he left the table with a very pretty woman. I have never had any intentions of risking an encounter in a whore house and have always avoided them. I declined every invitation, and boy were they persistent. A short while later, he returned, grabbed me by the arm, and said we must leave "right now!" We quickly got into a taxi, and I noticed many cops filling the street behind us. Luckily, we got back to Brownsville ok. I asked him what happened. He reluctantly told me the story. Apparently, the pretty girl was a man. After discovering this, he angrily got up and punched "her" in the face for not disclosing this before they were becoming intimate. Making out with a man was not something that Jim would want to brag about. I held back that story as a check on him a couple of times when he was about to do something stupid again.

Before moving on, I will tell a couple more funny stories involving my college friends who returned to Edna with me for the weekend. Richard and I had just returned home from a night of partying in Victoria. After getting out of the car, we noticed my childhood neighbor, Sandra, across the street. She invited us over to smoke a joint with her in the front yard. Before entering my parents' small house, a 3 BR/2BA ranch, I warned Richard that we needed to be very quiet, so we don't wake up my parents and younger brother and sister. I asked Richard to put his hand on my shoulder while I led us to my bedroom in the dark to avoid waking anyone.

Just about everywhere we stepped there was a scream and a soft spot.

My Dad rushed in and turned on the light to see what was happening. It turned out that my little sister was having a slumber party and about 8 girls were sleeping on the living room floor. Luckily, there were no injuries. Richard and I were laughing so hard that we could barely control ourselves. My sister later asked what was so funny. I did not tell her the full story until she got a little bit older. We still laugh about it today.

During another visit, my other friend Bob and I headed to Schulenberg for a dance. We shared a 6 pack, and he lit up a joint. Unfortunately, I got stopped for speeding shortly afterwards on an isolated country highway. We quickly rolled down the windows and hoped the car had aired out enough before stopping. Two policemen got out of the car. I got out and started talking to one. He asked me if I had been drinking and I said, "yes, I had a beer." Meanwhile, the other policeman started talking to Bob. He responded, "No sir, I have not been drinking." Bob had stacked up the full and empty cans along the inside of his door. The officer asked, "Do you mind if I look?" As he opened the door, all the beer cans landed on his feet. We all broke out laughing. The officers made us empty the other cans into the ditch. Luckily, we were in a different time and in a German part of Texas where everyone grew up drinking beer. Towns just south of there, including Edna, were largely protestant and had less tolerance for alcohol. You know, the ones who sneaked in through the back door of Jack's Liquor Store for their purchases and depart in secret. The officer just gave me a warning, and I was so thankful he did not smell the pot. Looking back, this could have had a worse outcome since marijuana was a felony with prison time, as was the case in the next story.

I ran into Candy Barr at a club in Victoria while with a college friend. I believe we were about 19 and she was in her mid-30's. She was still gorgeous and would later pose for Oui Magazine (a now defunct offshoot of Playboy) at age 40. At the time, I believe she was the oldest women to pose in a men's magazines. I Googled the issue while writing this book, and confirmed that yes, she was still hot. My friend immediately started to flirt with her. I did not tell him who she was but knew that the conversation could get interesting. At one point, Juanita said she just got out of prison and had also shot her husband. "He was an asshole and deserved it!" You should have seen the look on my friend's face as he pulled away. She never got indicted for shooting her husband. The prison sentence was for marijuana possession. That was the time I asked her why Jack Ruby shot Oswald. Nothing but a smile and a wink.

Like my dad, I had many crazy friends and we both did our best to keep them out of jail. There seems to be no shortage of crazy people in Texas. Dad liked to tell funny stories involving his best friend, Bedford, occurring during their single days. During a night out on the town, they had run out of beer money and were wondering what to do. Bedford walked over to the most dangerous looking guy in the bar. He bet the guy $20 dollars that he would not let him pull a hair out of his nose. If he succeeded, the guy won. Bedford pulled out some tweezers and started tugging on nose hairs, without pulling them out. As tears ran down the man's face, he finally grabbed his billfold and paid up. With beer at about 25¢ at that time, they were more than set for the night. On another occasion, they were at a high school football game at the stadium in Edna. It was common to bring a flask with whisky to the games. So, a few in the crowd were high by the end of the game. An out-of-town fan sitting next to Bedford was engaging him in bets but kept losing. After the game, while leaving the stadium, Bedford made one last wager. "I bet you $20 you can't knock me on my ass." The guy was angry about losing all his money that night and let him have it. As Bedford lay on the ground my dad said to himself: "Oh shit, we have a fight now." But Bedford got up grabbing for his billfold.

Ok, back to the serious part. I needed to find a research field. Knowing this is often a life-long direction, I kept searching for something that I loved. I gravitated toward heart research, specifically cardiac pathology, mainly due to a cardiac physiology lecture that caught my attention. I locked in on the structure-function association in heart enlargement disease (cardiac hypertrophy). Cardiac hypertrophy is enlargement of the heart primarily from growth of contracting muscle cells. This occurs from overloading conditions like hypertension and myocardial infarction where surviving heart muscle cells have to work harder to make up for the loss of contracting cells. One reason that heart diseases are so bad is largely because they often progress to debilitating health and eventual heart failure. **Heart failure** *is a syndrome, a common end point for heart diseases and is characterized by a dilated, dysfunctional heart and the likelihood of an earlier death.*

Note: when starting my career, we were not aware that there was also a form of non-dilated heart failure. More about this later.

In the 1970s, physiology was king and molecular biology was in its infancy. In a lecture during my first year of grad school, the professor discussed how the heart grows during cardiac hypertrophy. The accepted theory of parallel myocytes slipping past one another laterally made no sense. Imagine two rows of 5 pencils each slipping past one another laterally and becoming one row of 10 pencils. This mass would be thinner but larger in area. Theoretically, this could explain how changes in myocytes account for the dilated, relatively thin-walled ventricle observed in dilated heart failure. But how could cardiac myocytes slip past one another laterally when they are connected by side branches that would obstruct such movement? I believed that understanding cardiac remodeling in progression to dilated heart failure was key to understanding the syndrome. Most importantly, how could it be prevented or reversed? This was clearly an opportunity for new treatments. It was love at first sight. I finally found my calling! I later discovered that the cellular remodeling pattern of myocytes in dilated failing hearts was completely different than what was being taught. My work would go on to clarify how myocyte growth patterns account for anatomical changes in the heart under just about every known physiological and pathological condition.

When I was in grad school, there was no one in the Anatomy department who conducted heart research. A pulmonary researcher, Gerry Callas MD, PhD, who taught gross anatomy, was studying the structural effects of thyroid hormone dysfunction in the lungs. He agreed to become my PhD mentor, and I used his model to study cardiac hypertrophy from excessive thyroid hormones, primarily using electron microscopy to study structural changes in the heart. It proved to be a critical choice that would come full circle in my quest to discover a new treatment for heart failure. But I was largely on my own regarding mentoring. It was both a blessing and a curse. I became very independent from the beginning. New ideas were popping into my head almost daily. Luckily, I met a wonderful biochemist, Louis Sordahl, who studied cardiac hypertrophy. Importantly, he believed in me and introduced me to the broader literature on cardiac hypertrophy and heart failure. This was also helpful because biochemistry was an area of weakness for me.

Searching the research literature in those days before personal computers was much more labor-intensive. Each year a series of books called "Index Medicus" was published listing newly published papers. One could look up publications by topic or author. It was not clear how many years you had to go back to pick up most of the relevant literature. I usually stopped searching after turning up a few relevant reviews that seemed to be catching most of the same literature. But there were always surprises that were missing, and people reviewing your manuscript were often the authors of the missing citations (ouch!). Development of PubMed, a search engine for medical research literature, was a game changer. Today this is also used by savvy patients to ask more in depth questions of their physicians. Some of my physician friends in primary care tell me they sometimes run across a patient who seems to know more than they do about their disorder. AI now does that on steroids!

Med students often have nicknames for faculty. Some called Dr Callas the "Pigmy Dissector" because he was very short. Others called him the "Grenade Dissector" since he was known to tear through some important structures while looking for the one requested by the student. Texans are pretty good at coming up with nicknames. For instance, a short person may be called a "Sawed off runt." To me, that one is a bit overkill and on the cruel side. I am 6'1" and thought of it as a disadvantage every time I bumped my head. My grandfather was taller and had a different take: "Anyone can stoop, but not everyone can reach". After landing my first faculty position in Tampa years later, I had three women working in the lab that were all under 5' tall. I did a lot of reaching in those days.

There was another professor that med students liked to mess with. Lectures were in an old amphitheater with seating at a steep upward angle. One day, students grabbed this professor and began passing him around the audience, even turning him upside down. He just kept lecturing while they eventually placed him back on his feet on the other side of the auditorium. Maybe they thought he was a rock star.

Dr Callas was a wonderful man who taught me the art of electron microscopy. Early on, I began to realize that tissue artifacts likely tainted the interpretation of much of the data in the literature. This sent me on a quest to fine tune the methods and find corrections for artifacts so that my observations accurately reflected the underlying pathology. It turns out that the heart is so metabolically active that it must be fixed very rapidly to preserve intracellular structures accurately for detailed observation. It was clear to me that one had to rapidly perfuse the blood vessels and capillaries with a fixative to preserve underlying structures quickly for examination. Much of the literature related to structural alterations of cardiac myocytes in heart diseases was collected from heart tissues that were dropped into fixative. By the time the fixative penetrated the tissue, areas more distant from the surface were already undergoing degradation. So, many pathological observations reported in the literature were likely incorrect. This was compounded by the likelihood that diseased hearts were already stressed and may break down more rapidly during the lag period when hearts are removed and the fixative gradually diffuses into the tissue. How did I reach this conclusion? I saw this directly in my experiments by comparing areas of cardiac tissue that were well-perfused with fixative (capillaries were open and cleared of cells) to those that were not perfused (capillaries were closed and often obstructed with clotted blood cells). It is important to flush the capillaries with saline to clear the vessels of cells before perfusing with fixative. I never saw the reported "pathological" changes in well-perfused areas of hearts from hyperthyroid rats. They only occurred in areas that were not rapidly perfused, where tissue fixation occurred very slowly as the fixative diffused to deeper areas. These artifacts could also be observed in areas poorly perfused with fixative in control hearts but to a lesser extent, likely because those hearts were not stressed to begin with.

Did poor fixation affect our understanding of human cardiac pathology in heart failure? It certainly did! Several electron microscopic studies of immersion-fixed, failing human hearts showed dramatic myocyte changes, including severely damaged mitochondria, over-contracted myocytes, and disorganized contractile filaments. I knew these changes can also occur from poor fixation. And the common sense I learned growing up around people who could solve problems and fix things led me to a simple question. If such extensive lethal cellular changes are happening throughout the heart at this snapshot in time when the tissue was collected and placed in fixative, how were these people alive? Their hearts would have self-destructed in hours if this were going on extensively as reported. For me, it was another area of research needing clarification. It was clear that I had enough to keep me busy for the rest of my life. It also made me realize that some of the smartest people in the world can sometimes overlook or ignore simple things. I teach my students to question things that do not make sense.

You may remember that I started the book with joke about Texas grammar. When I submitted my first paper for publication, it was returned with some nasty comments about my grammar. One reviewer said that "I should find someone who actually spoke English" to help with the revision. It was my first realization that you can't write like you talk if your language is Texan. A professor recommended that I refresh my grammatical skills by reading Strunk and White. I did. Ever since then, I am annoyed if at least one reviewer does not comment that the paper is well-written. It only goes to show that some men are trainable. Good news for women. But I must admit that it took many years to clean up my speaking grammar. You can leave Texas, but Texas may never leave you.

I will close this chapter with another funny story. In addition to being a medical school faculty member, Dr Callas and his wife taught dancing on the side. Adagio was one of the dance styles. This is a form of ballet where the man sometimes threw the woman in the air and caught her. While I was in his lab, Dr Callas was approached by a casino in Las Vegas who wanted him to train a *nude* adagio act. He declined, of course. But I remember a discussion we had regarding "What could possibly go wrong?" Food for thought.

PART TWO

Scientific Discovery and Breakthrough

Chapter Five
Starting My Career as a Scientist

In 1978, I obtained a PhD in Anatomy at the University of Texas Medical Branch in Galveston after four years and launched a career in heart research. There was a small December graduation that I was unable to attend. I had promised my best friend that I would take his wedding pictures, and it was on my graduation day. Little did I know that I would finally make it to back to UTMB for graduation 45 years later and give the commencement speech as the 2013 Distinguished Alumnus. A lot happened in between.

I accepted a postdoctoral fellowship at Louisiana State University Medical School in New Orleans in the lab of Dr Fred Kasten in the Anatomy Department, one of the few anatomists doing heart research in the US. I learned the valuable technique of culturing heart muscle cells from neonatal rats. My non-scientist friends loved to look into the microscope and see these cells beating away. Funding for my position was part of a larger NIH training grant headquartered in the physiology department. In those days, research was largely siloed within a given discipline with little cross fertilization. When I gave a seminar on my structural heart research, I was peppered with only physiology and biochemistry questions by others with PhDs in those areas. I was much weaker in these areas having focused largely on anatomy and histopathology. It was a bit embarrassing for me. Importantly, it taught me that I needed to become more well-rounded. I returned the favor by peppering these physiologists-biochemists-pharmacologists with pathological questions during their seminars. They were equally embarrassed. As an indication of how they viewed my area of research, I was often asked to provide a "pretty picture" for their manuscripts. They didn't seem to think morphology was real science. I must admit that the early days of research in my chosen area made me feel at bit like the "Rodney Dangerfield of heart research", the comedian who joked about getting no respect, especially from his wife. But I knew they were wrong. There is a lesson here for young people who have good ideas but may get pushback because they are out of the mainstream. If you know in your heart that you are right, be persistent. I have learned that scientists who are original thinkers, those who don't jump on the latest bandwagon, often make the biggest marks in science. It's also ok to be the one who starts the bandwagon.

Dr Kasten was more of a pure basic scientist, while I was more interested in translational research, animal research that could lead to improvements in treatment of heart diseases in humans. Indeed, a major reason why I was not focused on medical practice was that a single big discovery could change medical practice from now on, possibly affecting millions going forward around the world. I wanted to be that guy. One could never have that effect by treating one patient at a time.

During my stint in New Orleans, I ended up with 3 good publications, all based on my pathophysiological ideas rather than Dr Kasten's more basic research goals. Thankfully, he was supportive of my excursions and the technical skills learned in his lab were extremely important. In one study, I teamed up with a physiologist doing heart research in dogs. The collaboration ended up producing the definitive paper on the **quantitative morphology** (morphometry) of normal dog heart(1). By that I mean, defining the normal composition of different cell types and internal organelles within those cells, especially contracting cardiac myocytes (e.g. 75% of heart tissue is occupied by myocytes). This contributed to our understanding of subtle differences between larger and smaller mammals, but also how strikingly similar they were. For example, we know that rats have a much faster heart rate, and hence, a higher metabolic rate than humans. The human heart beats at about 70 times per minute while a rat heart beats at about 400. We know that mitochondria provide the energy for contraction. Reflecting this physiological metabolic difference, contracting muscle cells (myocytes) in rats have about 35% mitochondria, the major energy provider, and humans about 25%. But the concentration of capillaries supplying the heart muscle and the size of contracting myocytes is identical in rats and humans.

I had to teach myself about morphometry to conduct this type of research. A researcher at Penn State, Larry McCallister was very helpful. He was a great guy and an up-and-coming star. But I later found that his department chair had a very detrimental effect on his career. It was notable that his chair always had multiple positions open, jobs that never seemed to get filled. I once interviewed there. One of the faculty I met with, clearly the superstar researcher in the department, had a telling message. The first thing out of his mouth was: "You don't want to be here." I later learned they finally got rid of this toxic chair. He moved to Hobart, Tasmania. A colleague commented that this was not far enough, another galaxy would have been better. This was one of my first lessons in how destructive toxic leadership can be for morale at an institution and the careers of promising young faculty members. Of course, this is a generic problem in academia and business alike.

I enjoyed my time in New Orleans and survived life as a young single man. It was quite a lifestyle change compared to Texas. First, Texas bars closed at midnight on weekdays and 1 AM on Saturdays at that time. So, they began to get crowded between 9 and 10 PM. The locals in New Orleans headed to bars at what I considered closing time. I remember dancing with my girlfriend while the sun was coming up overlooking Lake Pontchartrain, taking her home, and seeing that her parents, their friends, and relatives were still partying. My girlfriend was a beautiful redhead with green eyes, an artist who did some modeling, with a regular office day job. Her dad, who showed me his scar from open heart surgery, was from Spain and her mother from Nicaragua. Clearly, the late-night Latin culture thrives in New Orleans. Dance till you drop (dead of old age), seemed to be the mantra. When I left New Orleans for my next job at the University of Alabama Birmingham, I asked her to go with me. But she could not bear leaving New Orleans and her parents did not approve (I was 28 and she was 22). Her dad often said she was smarter than her brother who became a heart surgeon. But she decided not to attend college. It saddens me when someone gifted with a wonderful brain does not put it to use pursuing a meaningful career. In the end, I believe that was the important difference between us. The "Big Easy" can have an adverse effect on ambition. I could not see spending my life there as much as I fell in love with the city. It was just too much of a party place, even for me. At my core, I am basically a serious guy and a bit on the nerdy side. Importantly, I was not ready to marry at that time because I wanted to be more financially secure before taking that step.

After two years in New Orleans, I accepted a position at the University of Alabama Birmingham (UAB), a top 20 medical school. In going there, I turned down a faculty position in Tampa, a place I would rather have lived. But the one researcher in the world that I had the most in common with offered me a research faculty position in the Pathology department at UAB. Dr. Sandy Bishop was quickly becoming the most widely recognized expert in the field of cardiac pathology regarding heart enlargement diseases leading to heart failure. He was a DVM with a PhD in Pathology who ended up at a medical school. He had three other young veterinarians in his lab who were working on a PhD in pathology under his guidance. He convinced me to turn down the job in Tampa. We were both focused on the same problem. If a reliable method to accurately measure dimensions of contracting cardiac myocytes could be established, we could understand the underlying basis of important growth processes in heart diseases (later known as cardiac remodeling) and when they became abnormal. Using such a method, we could also determine if these highly differentiated cells are capable of cell division in adults. We could actually determine the number of contracting myocytes in a heart. This topic was gaining considerable attention due to research by Piero Anversa contradicting the accepted dogma that heart muscle cells in adults have lost their ability to divide. More about this later.

My tenure at UAB, from 1980-1982, was an exciting period. I saw, first-hand, what it takes to join the elite top 20 in NIH funding. Despite what other reservations you may have about George Wallace, he played a key role in making this happen. It seemed that everyone was on the same page at UAB, supporting each other, and all driving toward excellence. The editorial team for the journal *Hypertension*, a leading journal of the American Heart Association, was headquartered there at the time. Surviving a seminar there was a gauntlet through the needles eye of world-class researchers. Several from our group would go on to become President of the American Heart Association. It was the type of experience that I really needed in my early career.

Of course, Birmingham, AL, is not necessarily on the list of places most academics would choose. Among the important things I learned at UAB was what it takes to recruit and retain top scientists in a difficult place. The atmosphere at UAB at that time was amazing. Lessons there would later become important when I decided to go into administration, to become a department chair. Don't get me wrong, Birmingham is actually a beautiful city. But racial tension was still very high at that time. They had just integrated the bars. It appeared that blacks and whites largely just stared through each other as if they did not exist. Mixed-race couples were few and far between, with reactions ranging from disgust, having a death wish, or being accepted by those like me. Birmingham was the polar opposite of New Orleans, a city with a long history of racial mixing. I remember starting a conversation with a well-dressed black man in a Birmingham bar sitting next to me. He was clearly shocked to have a white man initiate a conversation with him. He was a successful lawyer in the black community and a very interesting guy. There was also prejudice against outsiders even if you were white. My boss Dr. Bishop, a white man from Vermont, could not become a member of the Birmingham Country Club because he did not have a generational local heritage. He seemed a bit annoyed that my girlfriend's father was a member, going back generations, and I ate there many times as a family guest. Apparently, this is a deep South thing. John C. Calhoun, a former US Senator and Vice President, had to be buried in Charleston's "strangers" cemetery rather than the one for locals from prominent families where his wife was buried.

While in Birmingham, the country band *Alabama* had rapidly built a strong local reputation. They were to become the house band at a new club opening downtown. I was looking forward to the grand opening. But Alabama never showed up. Their reputation exploded with a new album release. The rest is history. The movie Urban Cowboy came out at that time (1980) and country dancing exploded across the US. The movie centered around Gilley's Place in a suburb of Houston. It took me back to some great memories of that place while living in Texas. I remember being at Gilley's once on a Tuesday night and commenting on how dead it was. I bet there were only ~1,500 people there that night. You could see the other side of the room since the smoke was not so thick that night. With a 6,000-person capacity, it was listed in the Guiness Book of Records as the "World's largest honky-tonk." Thank you, John Travolta, for providing an accurate snapshot of where I grew up! The exciting nightlife provided a well-deserved reprieve for many who worked in the booming chemical and petroleum industries there.

There was also another annoying Deep South feature that was clearly different from Texas, the *requirement* to be "nice" to someone you dislike. I will provide an example. While visiting with a woman I knew from my apartment at a local nightclub, a passing "friend" stopped to chat with her. They both embraced and talked about how good it was to see each other. They went on and on, gushing for about half an hour. When the woman left, my neighbor said: "God, I hate that woman!" I thought: "What? You guys were so friendly". She responded: "I was just being *nice*." That trait was hard for me to get used to, not knowing if someone liked you or disliked you. I wasn't that good at reading between the lines. In Texas, if you don't like someone you just walk away or beat the shit out of them. You don't pretend to be nice.

I was never a good fit for Birmingham. I landed a major NIH grant. It was my ticket to leave. I called the Anatomy chair at the University of South Florida to see if he still had a position for me and my new grant. The position had been filled. But he went to the Dean who created one for me. It never hurts to ask. So, I moved to Tampa. The pathology chair at UAB warned me that going to a lower tier medical school would be the end of my career. He begged me to stay. I knew I would prove him wrong. I also missed the beach, and Florida Gulf Coast beaches are night and day superior to Galveston beaches (BTW South Texas beaches like Padre Island are very nice too). I eventually spent 10 years at the University of South Florida, 1982-92. Contrary to the prediction, it was a very successful period for me.

Upon arrival in Tampa, the first girl I dated was a scuba diver and immediately enrolled me in a scuba class. I had long wanted to try scuba diving. Our family vacationed at Camp Warnecke in New Braunfels, TX when I was a kid. It was on the crystal-clear, spring-fed Comal River. My uncle, Raymond (Mom's younger brother), who attended college near there would snorkel and spearfish, often going home with a nice fish dinner. So, I followed suit and loved snorkeling. Within a month after moving to Tampa, I was a certified diver. I did a lot of diving and spearfishing during the time I spent there. The biggest fish I got was an

80-pound amberjack. Unfortunately, I lost the picture. But I have included a pic of me with a nice catch of grouper **(Figure 5)**. I shot the amberjack in the head with a free shaft with no attached line and, boy, did he take me for a ride after I grabbed the spear. I was thoroughly exhausted when I got him on the boat. We dove several times a month from March until November, when the Gulf temperature dropped quite a bit, requiring a wet suit and the waves were often too high from storms. But we could go across the State in the other months and dive in the Gulf Stream off the coast of West Palm Beach where water temp was in the 70s, or drive to the Keys. You can literally dive year-round in FL! Drift diving in the Gulf Stream off the Atlantic coast was fun. One of the group divers had a rope attached to a surface buoy and the boat followed the drifting buoy to keep up with the dive party. Visibility was always excellent in the Atlantic, but you had to keep track of the guy with the buoy, or you may lose the group of divers and end up far from the boat when you surfaced. You could drift along with the current effortlessly watching all the coral and sea life go by. I got my first up close look at a moray eel, not something you would want to get bitten by. I absolutely loved diving with tropical fish. Of note, some of the women who dove with us were excellent at spearfishing. Yes, there were days when I was outdone by a woman. Here's to you Amy!

Most of the time, we dove off Clearwater Beach, where there was sometimes top to bottom visibility farther out. It was cool to look up from the bottom and see the boat on those days. But 20 feet was typical visibility and still very good. We would usually do 3 tanks a day. The first one was 70-80 feet deep, about 10 miles out. We used the Navy Dive Tables to avoid the bends (excessive nitrogen in the blood). After blowing off some nitrogen for an hour or two, we could do another shorter dive at that depth or go further in to about 60 feet and use a full tank safely (about 25-30 minutes). The final tank was usually at about 30 feet where nitrogen buildup was no longer an issue, and you could stay under as long as you had air. I dove with some locals who were very good divers and experts at spearfishing. After finding an area with detectible fish activity, usually associated with a rock pile/coral reef, we dove in. The idea was to stop when you saw the bottom and look for the fish. Never go to the bottom and approach the fish where they can easily spot you. After spotting the fish, you then proceeded to slowly descend on top of the unsuspecting biggest grouper or redfish you saw. Sometimes you could reload and spear another fish before they realized what was going on. But mostly you headed for the little caves in the coral rock piles where they fled to. If lucky, there were a few more fish to be had, sometimes with the help of a flashlight. In that area, the spiny lobster often got very big, unlike in the keys where they tended to be overhunted. I sometimes ended up with a couple lobsters between 6-10 pounds each. You were not supposed to spear the lobster. Proper technique was to sneak a wire loop behind them and snare them. Sometimes, you would see a group of antennae giving away their location, with the possibility of catching more than one. They are not very smart and often too curious for their own good. If skillful, you could slowly sneak a loop behind them and snare them while they stared at you.

For a while, I also dove with an architect who won the FL spearfishing title many times and eventually the World title in Australia. But I did not like the format of the spearfishing tournaments where you sometime found yourself shooting a fish that you would normally pass on and allow to grow up a bit more. You were graded with a score for each fish species resulting in a cumulative score. Hence, you may often shoot a smaller fish to make sure you have one in that category. The black grouper seemed more intelligent. It seemed they knew the range of your spear gun and would just stare at you until you got too close, then dart away. They might even swim up to a spear and inspect it as you reloaded. But they were long gone by the time you pulled the rubber bands back to shoot again. Red groupers were not as clever and easier prey. Unlike the black grouper, I often noticed they tended to have worms when cleaning them. Maybe they were not as healthy. Importantly, I had a freezer full of fish and lobster to make it through the short winter. Lobster and grouper dinner anyone?

On a very clear day while diving at 60 feet, about halfway back up to the boat, I noticed a large (~8 ft) bull shark about 25 feet away coming right at me. He was going side to side, obviously following the blood trail from the fish on my stringer. Rather than drop the fish, which would surely have sent him away from me, I pointed my spear gun at him. The intention was to poke him in the nose or drop my fish at the last minute if necessary. It was a bit scary. But he suddenly saw me and immediately darted away. I clearly scared the hell out of him! It seemed clear to me that sharks do not intentionally attack someone they recognize as potentially threatening. Otherwise, they could easily sneak up from behind and have a nice meal any time they wanted. So, I learned to respect sharks but not fear them. Thankfully, I never ran into a great white shark. I hear they are fearless. Occasionally, someone would catch a great white in the area we dove. They were there.

I remember another interesting experience playing mind games with a giant grouper. These fish are also called Jewfish, but that name is less popular for obvious reasons. I came upon one of these fish near a rock pile. They are known to be gentile giants. He was about twice my weight I estimated and about 10 feet from me. He was looking at me very curiously. I decided to see how close he would allow me to approach. When I got about 7 or 8 feet away, he started spreading his gills as if he were huffing and puffing to scare me away. I found that I could move my head back and forth about 6" in his direction and he would start and stop this behavior. He had a clear boundary defining his territory. After playing this game with me for a couple of minutes, he swam away, probably bored. Do fish have personalities? Occasionally I spotted large sea turtles and there were always tropical fish. During the winter, we also drove to Crystal River to snorkel with the manatees, who swam up these warm waters as Gulf temperatures dropped. It was fun to rub their bellies and see they clearly liked it. I really miss those days and being so close to nature below the water. I can't imagine my life without these diving experiences. I hope a reader of two who has often considered diving may try it.

Before ending this chapter, I would like to provide an example of how it can sometimes be challenging for innovative scientists to move forward with new ideas. Like everything else in life, the *status quo* can be a formidable obstacle. While sitting next to Sandy at a scientific session of the American Heart Association meeting in Dallas, we heard a presentation by Dr. Paul Simpson from the VA Hospital in San Francisco. He presented results from cultured heart muscle cells derived from newborn mice. After a few days, these cells start to beat and enlarge (hypertrophy). Sandy's response was that this model was too far removed from the reality of living mammals. He said it would not go anywhere. Having learned this model while in Dr Kasten's lab, I disagreed with his assumption, which turned out to be the *status quo* opinion at the time. Paul's talk was the last one of the day and I went up to the podium to meet him. We hit it off immediately and decided to go out that night. We ended up at the most popular club in Dallas on Ladies Night. For the first hour or so, we discussed science. Paul lamented about how difficult it was to publish his findings. Then Paul looked up and said: "Oh my God, I don't think I have ever seen so many beautiful women in one spot before." Paul, a tall handsome man, got up and started to dance and did not sit down for 2 hours. And, of course, I followed suite. We both loved to dance and had many other things in common. For instance, we both believed that marijuana should be legalized. At that time, there were cases across the US where some convicted of marijuana possession could spend more time in prison than convicted murderers. It is now legal in both our states, NY and CA. Interestingly, President Trump, not known for his stances on social justice issues, just declassified marijuana (12/18/2025) from Schedule 1 to 3, an important step toward decriminalization and possible national legalization. As you may know, drunks tend to speed and potheads tend to drive slowly. I agree that mixing both can get dangerous and moderation is important. This reminds me of an old joke. A pothead pulls up to a stop sign and does not move. Why? He was waiting for it to turn green.

Paul's work turned out to be very innovative and changed the field of cardiac hypertrophy-heart failure research forever. Paul eventually published several landmark papers in top journals. He showed that one could use cultured myocytes to dissect molecular signaling pathways. For example, one could put a neurohormone into a culture dish containing cardiac myocytes. This might trigger increased beating rate or enlargement of the cells, features that could be easily measured. One could then put various blocking agents for different molecules believed to be sequentially involved in the affected signaling pathways into the culture dish. Temporal progression in the activation of signaling molecules could also be examined more readily. Paul identified key signaling pathways involved in myocyte growth and contraction. This is very difficult to characterize in living models. For instance, if you inject a neurohormone into a mouse, it will not only target cardiac myocytes, but also brain, skeletal muscle, vascular smooth muscle cells and many other cell types and organs throughout the body. Not only does this trigger responses in these different cells, but they may also release many chemicals into the bloodstream with far-reaching effects. It is extremely difficult to determine what causes what. Paul's work led to major advances in our understanding of molecular signaling in the heart. To fully understand the impact of his work, nearly every grant submitted to NIH for the past few decades involving cardiac hypertrophy or heart failure research included cell culture experiments. If you failed to include those experiments, there is a good chance the grant would be sent back with a non-fundable score. After Paul published this amazing series of papers, he became one of the most famous cardiac molecular biologists in the world. It is more challenging to be a thought leader, rather than a follower, but it's worth it. Again, you make some interesting friends in science.

Chapter Six

Laying the Foundation for Research Success

> *Note: As I move forward with describing progression of my scientific findings toward my career goal of improving heart failure outcomes in patients, I will include brief lay summaries in text boxes for those who would rather not get into the details. So, you should be able to have a good idea of the progression of my research by just reading the text boxes.*

Shortly after setting up my own lab in Tampa, I realized that I had all the key information needed to solve the big question that Sandy Bishop and I had been pursuing, how to accurately measure cardiac myocyte size. Cardiac myocytes are essentially shaped like elliptical cylinders. They are the third largest somatic cell in the body after skeletal muscle cells and neurons. It is believed that skeletal muscle cells can be as long as the muscle. So, a skeletal muscle cell from the sartorius, a muscle in the front of your thigh, may be ~1.5 feet long in a tall person! A neuron can run from your spinal cord to the tip of your toe! Heart muscle cells are about 130 microns long on average. This is about the resolution of your naked eye. So, in a suspension of isolated cardiac myocytes, the tiny dots you see after shaking the tube and shining a bright light through it may be individual myocytes or clumps of a few myocytes. This is a big cell if you can see it with your naked eye! **Figure 8** shows a low power microscopic view of a high-quality preparation of isolated adult rat cardiac myocytes. They are rod-shaped, or more precisely elliptical cylinders. Within a given heart, their range in size is considerable, with the largest myocytes being about 15 times the size of the smallest. With such a wide range in size and every myocyte being a little different in shape from others, you can see how challenging it is to determine average myocyte size. It is important to determine the average myocyte volume since other parameters needed to calculate the number of myocytes in each heart are readily obtained. Essentially, dividing the total volume of all myocytes in each heart by the average myocyte volume provides the number of myocytes in that heart. And, as indicated earlier, this data could be used to confirm if cardiac myocytes can be stimulated to multiply in adult hearts. To many, this is the holy grail for treatment of heart diseases, coming up with a treatment to replace lost cardiac myocytes, after a heart attack for instance. Unfortunately, I am pretty sure this will never happen in my lifetime. There is a long list of problems that must be solved. For instance, if you inject isolated myocytes into an infarcted area of the heart, having them survive, connect to other myocytes, and align properly has been an impossible feat to attain to date. I am betting on heart transplants from pigs (xenotransplants) to win this race to human translation. It seems to me that solving the organ rejection problem with pigs is a more likely problem to be resolved. Researchers are also making remarkable progress in humanizing pigs (e.g. replacing pig genes with human genes) by gene manipulation to make them more resistant to human rejection. When these gene changes are done at the whole-body level, this could also provide an unlimited supply of other organs for transplants, like kidneys, livers, etc.

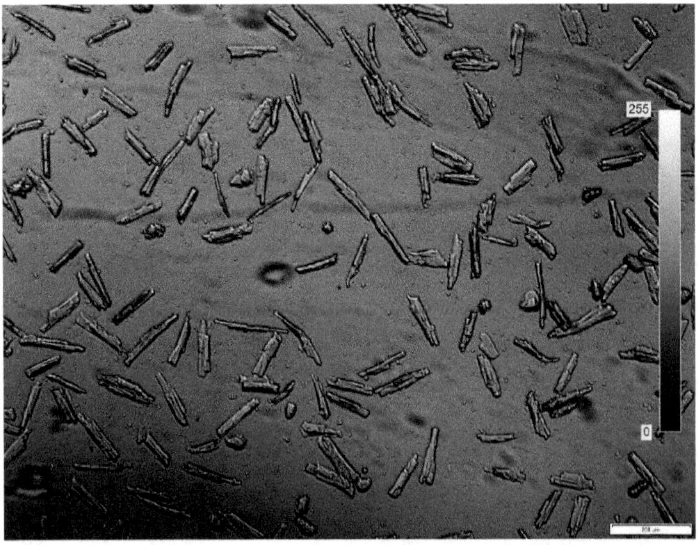

Figure 8: A preparation of isolated heart muscle cells from an adult rat heart. It is readily apparent that none of these rod-shaped cells are exactly alike. To further complicate measurement, myocytes are oval on cross-section, elliptical cylinders, not true cylinders. Since isolated myocytes settle on their flat side, this is a view of the major diameter with the minor diameter in the direction of view. Myocyte length can easily be

measured from these preps. Photo provided by Dr Yuan Huang from my research group.

Development of a technique to isolate individual, living, heart muscle cells from whole heart tissue was a breakthrough. Aside from accurate measurement of cell shape, contractile properties of individual cells can be determined. It is important to understand if changes in contractile properties of the heart are due to myocytes or if contraction is affected by changes in other structures surrounding myocytes, like excessive scar tissue (collagen) that may inhibit contraction. We made this distinction in one of our studies, showing that impaired contraction of the whole heart was due to excessive collagen around myocytes because their isolated myocytes had normal contractile ability(2). Such information would direct potential treatments to target inhibition of scar formation rather than myocyte contraction. The key step in isolating myocytes is to perfuse coronary blood vessels with an enzyme (collagenase) that breaks down the connective tissue binding the contracting heart cells together. It is a tricky process to literally tear the heart apart without killing the contracting myocytes. It took many investigators years to perfect. Our lab was the best at obtaining the highest concentration of viable myocytes from a given heart.

The information we sought was myocyte length, cross-sectional area, and volume. It is almost impossible to accurately measure myocyte length in whole hearts from tissue sections. Their shape is so irregular that it is uncertain if you are viewing the longest cell length within the plane of section being examined. The longest length may be outside the plane of section. Histology textbooks to this day still cite tissue sectioning data, while ignoring more accurate isolated myocyte data, providing an underestimation of the true average length of cardiac myocytes. Thankfully, medical students who memorize this relatively minor error are unlikely to endanger patients as a result. Unlike the case with tissue sections of whole heart tissue, one can easily measure maximum myocyte length by placing a drop of isolated myocytes onto a slide using a microscope **(Figure 8)** with a calibration marker (notice the 200 μm marker at the lower right corner of Figure 8). So, the problem with measuring cell length had been solved. The implementation of the Coulter

Channelyzer by Sandy Bishop to measure the size distribution of isolated cardiac myocyte volume provided the critically missing information. We use the average myocyte volume value for our work. But was it accurate? That needed to be determined. In leading up to this, we examined all the known factors that contribute to inaccuracies in the determination of myocyte dimensions by histological assessment. We determined the effects of tissue processing on myocyte dimensions and established a protocol that accurately preserves myocyte size without shrinkage or swelling. An approach was developed to correct for variation in sectioning angle in measuring myocyte cross-sectional area. If myocyte cross sections are not perfectly perpendicular to the long axis of myocytes, profiles will be more elongated because of the oblique sectioning angle leading to an overestimation of true cross-sectional area. Contracting myocytes also become shorter and thicker upon contraction. So, measurements need to be taken at the same degree of contraction or an appropriate correction provided. Fortunately, undamaged isolated myocytes in a preparation end up at the same resting, unloaded sarcomere length. Sarcomeres, containing bundles of sliding filaments, are the contractile units arranged in series along a myocyte. After conducting these technical studies, I realized that we had an approach to correct for all known artifacts in measuring myocyte dimensions using morphometric methods. So, we could take isolated myocytes from the same preparations and compare Coulter Channelyzer volumes to volumes obtained histologically. Myocyte volume = myocyte length X cross-sectional area. Cross-sectional area could be obtained from sectioned isolated myocytes. Corrections for all sources of error mentioned above were used to determine true average cross-sectional area. Unbelievably, the volume determination by both methods were virtually identical! This was a major technical advance in the field. Importantly, thousands of cells could be measured in seconds using the Coulter Channelyzer while obtaining volumes with the sectioning method mentioned above took weeks or months to complete a comparative study of just a few samples. One could collect isolated myocyte volume data by Coulter Channelyzer for an entire study of about 20 animals in a couple of hours! Importantly, this is also a completely objective method independent of human bias. We also collected myocyte cross-sectional area from whole hearts using tissue sections and corrected for all sources of error. Again, the values were identical to those from isolated myocytes. This was important because it confirmed the isolation procedure itself did not alter myocyte shape and affect measurements. Isolated myocytes were the same shape as myocytes from intact hearts.

Contracting heart muscle cells are cylindrical in shape and comprise 75% of the heart. It was logical to me that these cells were primarily responsible for changes in gross anatomy of the heart. My goal was to determine exactly how contracting muscle cells change their shape during normal and pathological growth. But there was no reliable method to precisely determine myocyte volume, length and cross-sectional area, the specific parameters needed. So, I had to tackle this problem first. It turned out to be a rather overwhelming task that many thought was impossible to do precisely. But I stuck with it. We eventually developed a method to isolate live, contracting myocytes from whole hearts, enabling accurate measurements of myocyte dimensions in addition to contractile activity. The method to determine myocyte dimensions was so accurate that results were literally absolute values that could not be improved upon. Armed with this method, I proceeded to characterize normal myocytes during growth and development and just about every rodent model of heart disease I could get my hands on.

Chapter Seven

Becoming "The Remodeling Guy" in the Age of Cardiac Remodeling

Armed with this incredibly rapid and accurate technique, our team proceeded to characterize changes in myocytes from just about every animal model I could get my hands on. We examined myocytes from rats, mice, cats, hamster, guinea pigs, and even humans! Yes, we were the first lab in the world to isolate high quality myocytes from human hearts obtained from the heart transplant program at the University of South Florida. Working primarily with Doug Schocken, Chief of Cardiology, and cardiac transplant surgeons Peter McKeown and Dimetri Novitsky, we published the first myocyte shape data from normal and failing human hearts in 1992. The paper published in *Circulation*, the world's leading heart journal, reported exactly what I suspected as a graduate student(3). The hearts were dilated due to elongation of myocytes by adding contractile units (sarcomeres) in series. A later study we conducted in an animal model demonstrated that elongation of myocytes alone could account for ALL of the heart dilation in failure(4). Myocyte slippage, the prevailing theory at the beginning of my career, was irrelevant and unnecessary to explain the cellular basis of heart dilation. I had answered the question that originally attracted me to heart research. But this was just the beginning.

> *I remember very clearly the first time I looked through a microscope at isolated myocytes from a human with dilated heart failure. I had never seen myocytes so long. We had just discovered the cellular basis for dilated heart failure in humans for the first time! Can you imagine being in my shoes at that moment?*

In 1986, we published the paper validating the Coulter Channelyzer method using isolated myocytes and whole heart tissues as described earlier(5). Although not the most cited paper from my lab, I believe it was the most important. It provided the key to unlock the truth about myocyte remodeling. We proceeded to provide data on myocyte volume, length, and cross-sectional area from many sources. We even examined cardiac myocytes from turtles. Unlike the situation in mammals, that study provided important insight into lower animals who retain the ability for cardiac myocyte division as adults. Focusing on mammals, the idea was to fully characterize the role of myocyte shape changes in normal growth and in heart diseases, especially heart enlargement diseases leading to heart failure. A reviewer once accused me of doing the same thing repeatedly, just changing the animal model. In other words, I was boring and uncreative. I responded, "Precisely. That is exactly what I was doing. I wanted to provide a *comprehensive understanding* of how the most important cells in the heart grow under normal and pathological conditions. I wanted to do this so well that it would stand the test of time and serve as a foundation for subsequent experiments investigating the molecular mechanisms of these changes." A unique feature of our cell sizing method was that it was so accurate and comprehensive that there was literally no room to improve data, the numbers were close to absolute values. We could even identify slight genetic differences between normal rats from different suppliers.

> *Using the precise myocyte sizing method developed in our lab, we assessed myocyte dimensions in many animal species and models of heart disease. We were also the first lab in the world to isolate high quality cardiac myocytes from humans when I worked with a heart transplant program. In dilated, relatively thin walled failing human hearts, the myocytes were long and skinny reflecting the gross anatomical changes. This disproved the prevailing theory that myocytes slip past one another in progression to dilated heart failure. Instead, it was an abnormal myocyte growth process that also led to functional problems for affected hearts. I knew we (it takes a team) were on to something important.*

Cardiac remodeling would later become the hottest field of cardiology, and many started referring to me as "The Remodeling Guy" because of the large body of valuable information I had already published and continued to provide. This was a very exciting time for me. I was being flown around the world to give talks in places like Paris, London, Prague, Heidelberg, Tokyo, Beijing, Amsterdam, Pisa, Munich, Athens, Cancun, Guangzhou, Wuhan, Cambridge (UK), Leiden (oldest university in Netherlands), Busan (S Korea) and many places in the US and Canada such a Toronto, Vancouver, Winnipeg, Ottawa, and Edmonton. This is clearly one of the perks when your scientific career is going well, free trips around the world to share your knowledge. Again, put yourself in my shoes and imagine how that felt.

They treated me particularly well in Beijing after giving my talks, assigning a personal car, driver, and young cardiologists in training to accompany me to places like the Great Wall, the Summer Palace, great restaurants, or just discussing science and life over beers. I have been to China many times and have seen the amazingly rapid development of this country up close. I remember on my first trip to Beijing that 8 of 10 lanes in front of the Forbidden Palace were for bicycles. Now, that is reversed. I was amazed by the number of cranes associated with all the high-rise buildings going up in many Chinese cities. I remember seeing many American couples in Guangzhou, there to adopt young girls who were not wanted during the one child policy. There was a Western pharmacy next to the traditional Chinese medicine pharmacy at the new hospital in Guangzhou they were showing off to our group of visiting scientists and cardiologists. There seemed to be a greater preference for traditional Chinese medicine at that time. I also noted some very sick people waiting to see a doctor in the hospital and later found they were among the original cases of bird flu. Whew! I dodged a bullet there. I also had a sleeping berth in an overnight train between two major cities in China. Now that train only takes a couple of hours. It was amazing to see the transformation of China up close. At these meetings, I was often one of a select group of Western experts receiving these prestigious invitations. I made many good friendships with world famous cardiologists and researchers at these meetings and enjoyed the local tourist sites the sponsors took our group to see. It is unbelievably rewarding from a personal standpoint to get to know these brilliant people from around the world and hear their personal stories.

I also had a ring side seat watching the fall of communism in Europe. I had a 6 month sabbatical in Munich in 1988 and later a NATO travel grant for several years to continue my collaborations. I remember visiting Germany just after the Berlin Wall fell in 1989 and seeing many East German Trabis in the ditches off the Autobahn. These poorly built two-stroke engine cars were no match for the high speeds. They were just left there, abandoned. Someone could have made a fortune on these cars today if they had thought of collecting them back then. I also visited Prague shortly after the fall of communism. There were no high scale restaurants yet but you could get drinks and meals for two people for under $10 then. On the bright side, the impoverished years of neglect in Prague left many historic buildings intact for preservation while they were being torn down and replaced with ugly modern buildings in the West. On the street cars of Prague I also realized that residents had not yet discovered deodorant.

I once had dinner in Cancun with Michael DeBakey (the famous heart surgeon at the Texas medical center in Houston) along with other invited speakers. He was 99 years old at the time and presented a talk about his latest artificial heart. I mentioned that I was living in Sioux Falls, SD at the time. He quickly shared a story of when he visited there 50 years earlier and was taken on a pheasant hunt. He remembered the name of the farmer who sponsored the trip. He also told a story from his childhood growing up in Louisiana. He and his brother used to try and shoot birds flying by with a 22-caliber rifle. They never hit anything (shotgun needed), but it was fun. Michael died later that year just before turning 100. What an honor to have spent personal time with such an amazing man, sitting right next to him, and getting to know him personally that evening.

As my reputation grew, I was invited to give talks around the world and became close friends with many of the world's leading experts in heart failure. I saw the rapid development of China and European developments after the fall of the Soviet Union up close. As a history buff, this was another exciting perk of being a scientist.

Work by William Grossman, then at Harvard, also provided a major driving force for my career. Bill is among the most famous heart failure cardiologists of his time. Based on human observations, Bill proposed that remodeling of cardiac myocyte shape underlies the observed anatomical changes from different loading conditions in the heart(6). With **pressure overload** (hypertension), the left ventricle is not dilated but the wall thickens to meet the increased workload and to reduce wall stress. He proposed that an increase in cardiac myocyte cross-sectional area accounted for this change. This cardiac adaptation, known as **concentric hypertrophy**, enables the left ventricle to pump a normal amount of blood at a higher pressure. This is like weightlifting where a muscle gets thicker, enabling it to lift more weight less stressfully. Unfortunately, chronic pressure overloading of the ventricles eventually may lead to scar formation, increased stiffness, and a non-dilated form of heart failure. This condition was not recognized in my early career and will be discussed later.

Heart failure should be considered a syndrome with various heart diseases, such as myocardial infarction (MI), hypertension, and other cardiomyopathies eventually leading to this outcome. The right ventricle, which pumps blood to the lungs, can also fail. For simplicity purposes, I will not go into right ventricular failure in this book. ***The standard definition of heart failure is when the left ventricle cannot produce enough blood flow to meet the metabolic needs of the body.*** *Of course, scientists being scientists, still debate proper wording of the definition.*

Back to science. Heart failure is one of the biggest healthcare problems in the world. The dilated failing heart is characterized anatomically by an increase in ventricular chamber diameter without a change in wall thickness, often described as a dilated, relatively thin-walled ventricle. Usually, we are talking about the left ventricle, which provides blood to the body. This increase in chamber diameter to wall thickness ratio leads to an increase in wall stress, which is especially important during the relaxation phase. We had proven that this increased chamber diameter/ wall thickness ratio is primarily due to an adverse change in myocyte shape. Myocytes become longer through addition of more contractile units (sarcomeres) in series without an increase in diameter. Heart tissue receives blood flow during the relaxation phase of the heart cycle. Heart contraction squeezes the capillaries and impedes flow during that phase of the cardiac cycle. Progressive increase in wall stress is a detrimental change, especially during the relaxation phase since blood flow through the capillaries is reduced. This reduces oxygen delivery to contracting myocytes, leading to impaired nutrition and possibly cell death. For example, hypertension leads to increased wall stress in the deeper layers of the left ventricle during the relaxation phase and is associated with poor blood flow, myocyte loss, and increased deposition of scar tissue. As this myocyte remodeling process continues, patients get progressively sicker and may eventually die from heart failure.

The other part of Grossman's theory involved **volume overloading** of the heart, leading to what is known as **eccentric hypertrophy**. Volume overloading leads to a proportional change in chamber diameter and wall thickness as an adaptive growth response to normalize wall stress. The heart can be subjected to an increase in volume overloading by several conditions, such as a leaky heart valve, allowing some blood to leak back into the ventricle after it is pumped out. This forces the ventricle to not only pump the new blood being delivered but also some of the previously pumped blood that had regurgitated back into the ventricle. A slow increase in volume loading also occurs as we grow up and the body and heart get bigger. This physiological form of heart growth was also proposed to result from a proportional increase in chamber diameter/wall thickness due to proportional growth of myocyte length and cross-sectional area. Our lab later confirmed this was true(7). Refer to **Figure 9** to better understand the discussion of cardiac myocyte remodeling in this section. Note that I will use myocyte cross-sectional area and diameter interchangeably. They are both measures of myocyte transverse shape with the former being two dimensional and the latter one dimensional. Myocyte diameter correlates to wall thickness more directly since both are unidimensional parameters.

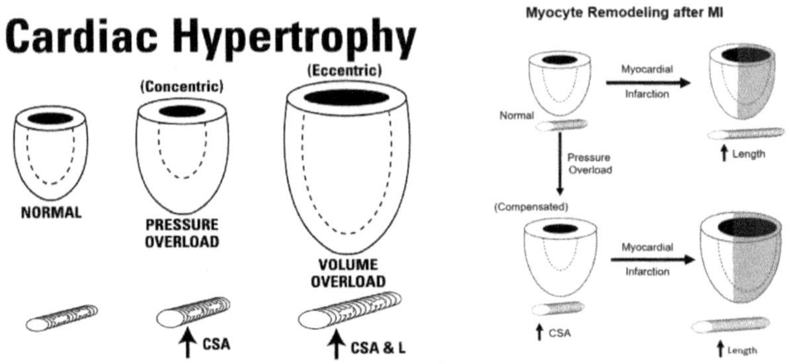

Figure 9: Myocyte remodeling in pressure and volume overloading (left). Myocyte remodeling after **myocardial infarction (MI)** with and without prior hypertension (right). Grey area is MI. White area is viable, non-infarcted myocardium. Note that myocytes are essentially shaped like elliptical cylinders.

There were no myocyte remodeling data to support Bill's theory of concentric and eccentric remodeling of the heart. As a graduate student, I became hooked on the idea of proving whether his theory was valid or not. It made a lot of sense to me. But I also wanted to discover how myocytes remodel in dilated heart failure, which was clearly not a beneficial adaptive cellular response to normalize wall stress as was the case in volume and pressure overloading. The response to dilated heart failure leads to increased wall stress and all the clinical problems that creates. This is because dilation of the heart is not compensated by an increase in wall thickness, so wall stress is not reduced as in eccentric hypertrophy as outlined above.

We convincingly demonstrated that anatomical changes in heart anatomy are reflected at the myocyte shape level. Thickening of the left ventricular wall, as occurs in high blood pressure, is due to thickening of myocytes. Dilatation of the heart, as may occur with a leaky heart valve, is due to lengthening of myocytes by adding new contractile units in series.

T4 (tetra-iodo thyronine) is considered the precursor to the more active form of the hormone **T3** (tri-iodo thyronine). These are very simple but powerful hormones, comprised of two tyrosine amino acids with 3 (T3) or 4 (T4) iodine molecules attached. To activate T4, an enzyme removes an iodine in a specific location on T4, converting it to T3. Another enzyme removes an iodine at a different location to deactivate T4 to **rT3** (reverse T3). The thyroid gland secretes much more T4 than T3. So, much of the activation of thyroid hormones, converting T4 to T3, occurs in peripheral tissue. This occurs mostly in the liver and kidneys, but the brain and heart also play a key role in regulating thyroid hormones within those organs.

It would be helpful to provide some basic information about thyroid hormone physiology before going deeper into thyroid hormone regulation of the heart and its role in heart failure. To understand why thyroid hormones matter in heart failure, imagine your heart as an engine. In heart failure, the engine is struggling. Traditional treatments try to reduce the workload, to put less stress on the heart (e.g. avoid driving uphill). But what if we could tune up the engine itself? That's what thyroid hormones do, optimize heart muscle cell function at the molecular level. Thyroid hormones regulate heart rate, generation of contractile force, and coronary blood flow. Obviously, normal thyroid hormone levels are important in optimum functioning of the heart.

I will provide more background on thyroid hormones since my research was moving in that direction. T3 and T4 are thyroid hormones. T4, the largely inactive precursor of T3, has 4 iodine molecules while T3, the active form of thyroid hormone, has 3 iodines. An enzyme in peripheral organs removes an iodine from T4 to activate it to T3. The thyroid gland secretes mostly T4 while T3 is largely created peripherally, primarily by conversion in the liver and kidneys but also in the heart.

There is an intricate interaction between the hypothalamus of the brain, the pituitary gland at the base of the brain, the thyroid gland in the neck, and target organs throughout the body. When these organs are interacting properly, the body maintains homeostasis of metabolic processes throughout the body. **Figure 10** summarizes the interaction of these organs to regulate this process. The hypothalamus releases **TRH** (Thyroid Releasing Hormone), which stimulated the pituitary to release **TSH** (Thyroid Stimulating Hormone). TSH triggers release of the thyroid hormones T3 and T4 from the thyroid gland into the blood where they travel to target organs throughout the body. When T3 and T4 levels are too low, hypothyroidism occurs and metabolism throughout the body slows down. This produces symptoms such as being overly tired and sleepy. To correct this, as indicated by the green arrows, low serum T3 and T4 stimulate the pituitary to secrete TSH, which triggers release of more T3 and T4 from the thyroid gland. Low thyroid hormone levels also stimulate the hypothalamus to release more TRH, which triggers increased release of TSH from the pituitary. If T3 and T4 are too high, release of TSH and TRH are inhibited, which leads to reduced secretion of thyroid hormones into the blood by the thyroid gland (red arrows). Note: fT3 and fT4 are the free forms of the hormones and only last about a minute in the blood before being degraded. The vast amount of thyroid hormones, over 99%, are bound to proteins in the blood where they are inactive in the bound form. When released from these proteins, T3 and T4 are in their activated forms (fT3 and fT4), with fT3 being far more active than fT4.

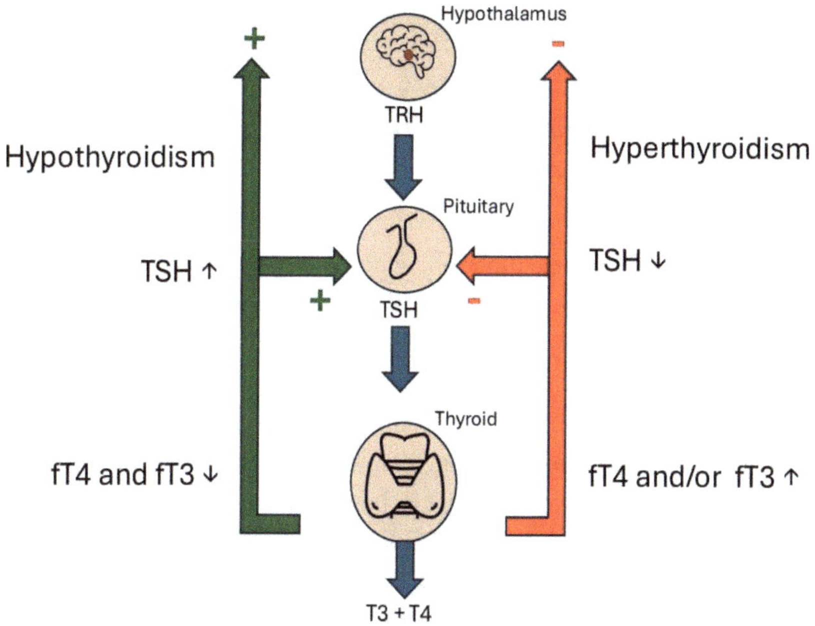

Figure 10. Interaction of hypothalamus, pituitary, and thyroid gland showing positive (green) and negative (red) feedback loops to maintain homeostasis of thyroid levels in the body.

After validating our cell sizing technique, we examined how myocytes remodel when thyroid hormones are too low (hypothyroidism) or too high (hyperthyroidism)(2, 8-10). Hyperthyroidism increases metabolic rate, leading to an increased amount of blood being pumped from the left ventricle to meet the increased energy needs of the body. Although this produces a more physiological increase in cardiac mass, it resembles non-stop exercise, should be considered pathological, and certainly merits corrective treatment. This condition is more obvious and readily diagnosable than low thyroid conditions. Using a rat model, we confirmed that hyperthyroidism leads to a proportional increase in chamber diameter and wall thickness. This is reflected at the myocyte level by a proportional increase in myocyte length and diameter. We showed how rapidly this growth process occurs and how quickly it reverses after discontinuing excessive thyroid hormone administration(8).

The hypothalamus and pituitary gland in the brain, along with the thyroid gland interact with one another to regulate normal levels of thyroid hormones in the blood to maintain metabolic homeostasis in the body.

Low thyroid hormone function, hypothyroidism, was more interesting to examine because animal and clinical studies reported this condition can eventually lead to dilated heart failure even though it is not of such urgent concern as hyperthyroidism. Because thyroid hormone treatments have been available for about 100 years, I realized that no one had ever *proven* that chronic hypothyroidism alone can lead to dilated heart failure. And, of course, how myocytes remodel in hypothyroidism was unknown. We examined temporal progression of hypothyroidism for up to one year, and yes, it eventually led to dilated heart failure as reported in a few human case reports over the years. Initially, hypothyroidism leads to reduced cardiac mass because of lower metabolism and reduced **cardiac output** (CO, the volume of blood the heart pumps per minute). The heart is not working as much and like skeletal muscle, it atrophies. This is associated with a reduction in wall thickness, which correlated with reduced blood pressure and a predictable reduction in myocyte cross-sectional area(9). But, over time, these hearts began to dilate due to excessive lengthening of myocytes(10). This is a clear heart failure **phenotype** (e.g. an observable trait that can be seen or measured, like body weight or height) and eventually these animals got very sick and were dying of heart failure-like symptoms. Despite lower systemic blood pressure, wall stress was actually higher due to reduced wall thickness and increased ventricular dilation. This was reflected by a reduction in myocyte diameter and an increase in myocyte length. This turned out to be a critical observation because growing evidence suggested that patients with dilated heart failure have a higher incidence of hypothyroidism and borderline low serum thyroid hormones. Although borderline low thyroid hormones in patients with heart failure are not generally treated, many recent clinical studies show these patients die sooner. This raises the question: "**How much of the patients heart failure symptomology is due to the disease (e.g.**

myocardial infarction, diabetic cardiomyopathy, etc.) or to a treatable low thyroid condition." More about this later.

We examined several rat models of hypertension and confirmed that the associated wall thickening was due to an increase in myocyte diameter **(Figure 9,** left)(11). To examine volume overloading more extensively, we developed a reversible rat model by surgically creating a shunt (connection) from the abdominal aorta to the parallel vena cava (large vein) bypassing the capillary bed. From a hand surgeon, my graduate student, Liu Zhi (now a pathologist at NYU in Manhattan), learned how to connect a side branch of the vena cava to the aorta, forming an **AVF** (Arteriovenous fistula). So, a considerable amount of blood would flow from the aorta across to the inferior vena cava and back to the heart without oxygenating the lower extremity. This caused the heart to pump twice the normal amount of blood. In later studies, we placed a metal collapsible ring around the shunt to reverse the overload and study regression of the associated eccentric hypertrophy. The volume overload from this model was so large that the heart doubled in size over a few months. As expected, we confirmed that cardiac myocytes had a proportional increase in myocyte length and width **(Figure 9,** left)(12). With this model and the hypertension models, we had proven Bill Grossman's theory(13). To this day, there is still much to learn about the process by which contracting myocytes add more contractile units in parallel (e.g. increased myocyte diameter or cross-sectional area) and in series (increased myocyte length). It is easier to envision how contractile filaments could be added to the peripheral edge of sarcomeres while continuously contracting during hypertrophy. But it is difficult to imagine how sarcomeres could be inserted in between myocyte sarcomeres while they are contracting. We tried to answer this question using electron microscopy realizing that one sarcomere per day was being added, in series, during the initial period of the AVF overload. But there was no smoking gun. Apparently, the process occurs very quickly.

***Hypothyroidism**, or low thyroid hormone levels in the blood, was long believed to cause dilated heart failure based primarily on individual case reports. But this slow process was never definitively demonstrated in a human study because thyroid hormones have been available for ~100 years and it would be unethical to withhold treatment from these patients. In a rat model of hypothyroidism, we clearly demonstrated for the first time that chronic hypothyroidism alone eventually leads to dilated heart failure. I say alone because humans often have a complex mixture of adverse things going on at the same time, such as atherosclerosis, diabetes, cardiac calcifications, hypertension, etc. This can make it more difficult to determine cause and effect. We also showed that increased thyroid hormone levels, **hyperthyroidism**, leads to remodeling of heart muscle cells like that of exercise. But as non-stop exercise 24/7 over time would clearly not be good, excess thyroid hormones can also lead to a type of heart failure. Like other hormones, the sweet spot is in the middle (Goldilocks).*

What about reversal of heart mass in a large volume overloaded heart after removing the overloading stimulus? What happens at the myocyte level during regression of heart mass? We had no idea. One possibility was that myocytes could remove series sarcomeres to reduce chamber dilatation. But this sounded impossible. How could this happen while myocytes were actively contracting. For this experiment, it was important to use a model producing very large hearts in case the reversal process was incomplete. Two months after creating the surgical AV fistula, a second surgery was done to crimp the ring around the fistula to close the shunt and stop volume overloading, which led to a rapid reduction in heart size.

*I remember looking through the microscope for the first time at isolated myocytes from these severely enlarged and dilated hearts after being unloaded to study reversal of the hypertrophy. **Whatever we observed, we knew it would be the first such observation in history.** It was another "aha" moment demonstrating a growth process that we thought was physically impossible. In this experiment, we* *showed for the first time that heart muscle cells had the ability to remove sarcomeres in series as the heart returns to a more normal size after reversing the overload. To this day, we have no idea how this happens while the heart is continuously contracting(14). But it certainly leaves an interesting avenue of exploration by future scientists who choose to walk in my shoes.*

Rats subjected to the AVF model of volume overloading showed no signs of heart failure after 6 months despite a doubling of heart size. This type of overloading, where the growth pattern resembles physiological growth during development, was well-tolerated, perhaps not so surprising. **But an important question remained. Does the *dilated failing* heart have this ability to remove sarcomeres in series?** If not, prospects for optimal treatment of dilated heart failure at late stages do not look good for patients. I will answer this question while also introducing a breakthrough in treatment of cardiac hypertrophy-heart failure that was occurring at this time. A new class of drugs was one of the most important advances in the care of heart failure patients and I was not only a major participant but had a ring side seat by showing the world how myocyte remodeling played a critical role in the process.

Prior to the 1980s, there were no drugs for dilated heart failure that enabled patients to live longer. Some drugs, like diuretics, could stimulate urination to remove some of the excessive amount of fluid volume in the circulation that was overloading the heart. Then, Janice and Mark Pfeffer from Harvard received an experimental drug called captopril. This drug was an **ACE inhibitor** (angiotensin converting enzyme inhibitor), which targets the **RAAS** (renin-angiotensin-aldosterone system). Scientists had determined that this system, which increases blood pressure and blood volume, was overactive in heart diseases leading to heart failure and suspected inhibitors of various components of this pathway would improve outcomes in these patients. Janice was a physiologist who tested captopril in two rat models of heart failure: a rat model of surgically induced MI and the **SHR** (Spontaneously Hypertensive Rat), a genetic model of hypertension(15, 16). Janice found that not only did heart function improve with the ACE inhibitor, but the rats lived longer. A key observation was that the extent of heart enlargement was significantly reduced. In parallel experiments, SHR rats treated with a vasodilator normalized elevated blood pressure but treatment did not affect heart weight or reduce mortality.

Because my work overlapped with that of Janice, I often ended up with a poster near hers at American Heart Association meetings. Janice, Mark and I had many interesting conversations. Mark is a heart failure cardiologist who led the first clinical trials with captopril in patients with heart failure. Janice's rat work, described above, set the stage for the human trials. They were a truly amazing husband-wife research team. The first human trial results, published in the New England Journal of Medicine, demonstrated the biggest breakthrough in heart failure treatment ever to that point(17). As shown in the animal experiments, captopril treated patients had better heart function, lived longer, and showed reduced heart mass (reverse remodeling or prevention of hypertrophy with early treatment). Improvement in patient outcome was clearly connected to improved cardiac anatomy. Technical advances in echocardiography and **MRI** (Magnetic Resonance Imaging) also improved the ability of cardiologists to better diagnose and follow progress of treatment. This study launched the era of cardiac remodeling in heart failure, and I was the guy who was best prepared to show specifically how heart muscle cells remodeled from these drugs targeting the RAAS.

Prior to the 1980s, there were drugs that could reduce symptoms from heart failure but none that improved survival. Mark and Janice Pfeffer showed that a new class of drugs, ACE inhibitors, not only reduced symptoms but enabled animals (Janice) and humans (Mark) to live longer with heart failure. A critical observation was that improved survival was linked to attenuation of adverse changes in heart anatomy. The age of Cardiac Remodeling began, and my work moved from the shadows to the forefront.

At this time (late 1970s-1980s), heart researchers worked predominately in "silos". Physiologists did physiology, biochemists did biochemistry, and pathologists did pathology. A few papers included physiology and biochemistry, but they were uncommon. At that time, I was largely looking at cardiac structural changes, mostly myocyte remodeling, while providing little additional insight into the associated changes with regard to physiology or biochemistry. Molecular biology was in its infancy. At an American Heart Association Meeting, I had a poster next to Heinz Gerd-Zimmer, a well-known physiologist from Munich, Germany. We both examined hyperthyroid rats treated with a Beta-adrenergic blocker (an antagonist that blocks signaling through a cell receptor). Like drugs targeting the overactivated RAAS system discussed earlier, we had learned over the years that adrenaline like hormones targeting the Beta-adrenergic system were overactivated in patients with heart failure. This overstimulation made sense as an attempt to help the failing heart maintain normal function. But on a long-term basis, it was becoming clear that this increase in stress due to activation of neurohormones was bad for the heart. This was a landmark concept in heart failure research known as the Neurohormonal Hypothesis. My good friend Jay Cohn, chief of cardiology at the University of Minnesota for 35 years, was a key player in this field that preceded the era of cardiac remodeling. Jay's group also led to major advances in designing and conducting clinical trials in cardiology, making them more scientifically sound. Many consider Jay *the* most important expert in heart failure that ever lived. He published an autobiography in 2022 at age 92. Jay once tried to recruit me, but the Chair of Medicine at the time would not allow tenured PhDs in this clinical department.

In 1988, I did a sabbatical with Dr. Heinz-Gerd Zimmer in Munich to incorporate state-of-the-art collection of heart function data from rats using new high-tech catheters. We examined structure-function in rats after a heart attack, combining our techniques. This was the first study to show that progressive dilatation in heart failure, characterized by a dilated, thin-walled ventricle in the undamaged part of the heart, was due to a maladaptive myocyte shape change. We would later show the same change occurs in humans with dilated heart failure. The collaboration with Dr.

Zimmer was unusual at the time where investigators from different disciplines typically worked in silos. Physiologists, biochemists, pathologists, and pharmacologists generally published in journals of their disciplines. More recently, they collaborate. This has led to publications that are more comprehensive- a good thing.

Back to Heinz, who was working with a company in Houston that revolutionized measurements of blood pressure and heart function in rats and mice using high tech miniature electronic catheters. In his adjacent poster at an AHA meeting, he reported left and right ventricular functional changes in hyperthyroid rats treated with a beta blocker. In my lab, we examined remodeling of left and right ventricular cardiac myocyte remodeling in the same model. There was such a good correlation of our results, it appeared that structural and functional data came from the same rats. I decided to do a sabbatical in Munich so we could cross-train each other in our methods. Since Heinz was an expert in surgically creating MI's in rats, we examined left and right ventricular physiology and myocyte remodeling in that model. The publication marked the first time that state-of-the-art ventricular function and myocyte remodeling data from the same animals were reported. This study was the first to show what I had long suspected. In heart failure, characterized by a dilated relatively thin-walled ventricle, an abnormal change in myocyte shape was responsible(18). In humans after a heart attack (MI), we know that the undamaged part of the left ventricle increases in area as the heart gradually dilates until the patients develop severe heart failure. During this progression, there is no change in wall thickness in the non-infarcted viable myocardium. The rat MI study with Zimmer showed that myocytes increase in length without a change in width during the progression to dilated heart failure (e.g. long skinny myocytes)(18). In essence, we showed that maladaptive growth of cardiac myocytes is an important change in progression to dilated heart failure. This was a major discovery that brought important attention to the development of potential drugs targeting this maladaptive cell growth process. Drug companies started contacting me to test their promising, newly developed drugs.

It was very important to demonstrate that human hearts respond the same way as rats during progression to dilated failure. Fortunately, our lab started collecting fresh tissue samples from failing hearts supplied by the heart transplant program at the University of South Florida at this time. With our considerable experience in isolating myocytes from many species and rodent models of heart disease, we set out to do the same in humans. Fortunately, a medical student, Scott Kellerman, took a year off from med school to make this happen. It was difficult work leading to many sleepless nights. His beeper would go off, usually early Saturday or Sunday morning and he would rush to Tampa General Hospital to collect large pieces of hearts, put them on ice, and rush to our lab across town. Donors were typically gun-shot victims or involved in fatal collisions. We called motorcycles "donor-cycles" because of their frequent contributions. Kellerman was a dedicated med student determined to perfect the isolated myocyte method in humans, and he did. When we got the first myocytes from a patient with ischemic cardiomyopathy (triggered by an earlier heart attack), we looked through the microscope in amazement at what we saw. It was one of those scientific moments you never forget. In the case of our rat MI data, cells were collected two months after the MI. In the case of human patients, hearts had been progressively dilating for many years after the MI in most cases. They were much more advanced than in our rat experiments. The human myocytes were unbelievably long but also had a normal width, as expected **(Figure 9,** right**)**. In some cases, the width was also increased. In those cases, we told the cardiologist that the patient likely had hypertension prior to the MI and progression to dilated heart failure **(Figure 9,** right**)**. Sometimes they said no. But we realized that heart attacks often lower blood pressure because the left ventricle is no longer healthy enough to generate excessive pressure. Sure enough, a check of the patient's prior history always confirmed our myocyte shape data. The cardiologists were amazed that we could describe the patient's history from myocyte shape.

Over the next few years, we collected additional myocytes from patients with the two main types of dilated heart failure, ischemic cardiomyopathy (MI) and idiopathic dilated cardiomyopathy (unknown cause, but not ischemic). In every case but one, the myocytes were always very long and skinny. **Early on there was one exception that**

threatened my theory of myocyte remodeling in dilated heart failure. In this case a patient diagnosed with dilated heart failure had perfectly normal shaped myocytes. This made no sense and we asked the pathologist on our team to investigate further. When he examined tissue sections from this patient's heart, a misdiagnosis was confirmed, to our relief. The patient did not have *chronic* dilated heart failure. Instead, this heart was dilated and failing due to an acute infection. The diagnosis was changed to acute myocarditis. Tissue sections of this heart showed significant infiltration of inflammatory cells (white blood cells), confirming the infection. The infected heart was too weak to accommodate blood flowing through it. Weakened myocytes did not have time to remodel but instead were stretched to their limit giving the appearance of dilated heart failure. Transplants are not generally made in this situation since there is a risk of infecting the transplanted heart and fighting the infection would be the proper path to take. In the heat of the moment in trying to save a dying patient, it's easy to see how this could happen.

Although subsequent patients continued to show the expected changes in dilated heart failure (e.g. long skinny myocytes), we did not get a non-failing, control heart for quite a while. Those hearts are transplanted, of course. I was optimistic we would eventually get a few control hearts. The issue of proper controls in heart transplant research is problematic because the good hearts get transplanted. Most non-failing potential donor hearts that are rejected for transplant have adverse conditions, like excessive coronary artery disease, age, or hypertension and associated wall thickening, etc. Of course, we had already shown in many other studies that average myocyte size is similar across mammalian species. But it made me a bit nervous waiting for confirmation of normal myocyte size in humans. Then, along came an unfortunate 21-year-old male. It was anticipated that his heart would be transplanted. But he was so high on amphetamines and cocaine they could not stabilize his heart. I am not sure about the rules of Russian Roulette but presume he lost. His myocytes were completely normal, exactly like we observed in every other normal adult mammalian heart. We eventually collected a few more non-failing controls, although wall thickening from hypertension was common. Importantly, we had proven that an abnormal change in myocytes shape was, indeed, responsible for the anatomical observation of a dilated, relatively thin-walled ventricle. The first report from patients with ischemic cardiomyopathy appeared in *Circulation,* the world's leading heart journal (3). We later submitted a follow-up paper to *Circulation* demonstrating that patients with idiopathic dilated cardiomyopathy, the other main type of dilated heart failure, had the same myocyte shape change. It was rejected with the editor stating, "this is the same thing you reported for ischemic cardiomyopathy." We disagreed, of course, arguing that confirming the same common cellular mechanism occurred in both types of dilated failure was important new information. You win some and you lose some.

Our studies of isolated myocytes from many mammalian species showed something else of interest. Not only was average myocyte size similar in all adult mammals studied, but the specific length to width ratio was also the same, ~8:1(13). There seemed to be something magical about this shape. In dilated heart failure, the myocyte length to width ratio often increased to 10:1 or more in humans. A few years after we published data showing that myocyte shape changes account for the dilated, thin-walled ventricle in heart failure, I got a call from a young investigator at Harvard. Kevin (Kit) Parker was excitingly saying: "I proved your theory, I proved your theory!" I asked him what he meant. Kit had been growing cardiac myocytes in culture while inducing different shapes using a molding system. He could grow star-shaped myocytes and myocytes of different length to width ratios. Kit studied the effects of myocyte shape on electrical properties involved in contraction. He found that myocytes of normal shape had optimum electrical activity but those with the elongated heart failure shape produced abnormal electrical activity (arrhythmias). I told Kit that he did not prove my theory but offered a significant advance in our knowledge. We know that patients with cardiac hypertrophy progressing to heart failure have an increased incidence of arrhythmias and sudden death. What Kevin showed was that the change in myocyte shape observed in dilated heart failure itself *promotes* arrhythmias(19). Myocyte shape is important not just in remodeling and its effect on wall stress but in maintaining optimal myocyte function! It has been fun to follow Kits career. He is a big guy and was a paratrooper in the US Army Reserve, serving 2 rotations in Afghanistan. He shared some very interesting war stories during his visit to my lab. He reached the rank of Lt Colonel. In between raids, he often reviewed grants for different agencies. His scientific career has really taken off since I met him. You meet some interesting people in science. I have also found that medical students often have an interesting early life. For instance, a medical student I had in Tampa was previously a high wire trapeze artist in the Barnum and Baily Circus.

I still needed to confirm the underlying myocyte shape changes associated with the Pfeffer's findings using drug-treated rat models of heart failure. Before embarking on rat studies of MI examining the potential myocyte changes resulting from breakthrough drugs like ACE inhibitors or Beta-blockers, I was interested in finding a rat model of heart failure with fewer limitations. The increase in heart mass after a MI is rather modest due to the loss of many heavier contracting myocytes replaced by lighter scar tissue in the infarct. Nonetheless, the left ventricle does get bigger after an MI and myocyte lengthening proceeds in the undamaged, non-infarcted tissue **(Figure 9,** right). But it would take about a year to simulate the extent of myocyte lengthening in rats that we saw in end-stage humans with ischemic cardiomyopathy at the time of transplant. Too many rats would also likely die of sudden death before scheduling terminal experiments if we waited too long. So, there was a good chance these experiments would take years to complete. The MI model lends itself better to prevention studies, initiating treatments at the time of infarction, as would be done clinically. We were most interested in a rat model to study the possibility of reverse remodeling with drug treatment.

We showed that normal cardiac muscle cells from mammals have a consistent length/width ratio of about 8:1. But in dilated heart failure the myocytes become long and skinny with a length/width ratio over 10:1. Kit Parker from Harvard showed that myocytes with a normal ratio have optimum electrical properties. But myocytes with a higher length/width ratio, as found in heart failure, have disrupted electrical activity. This provides another good reason why individuals with dilated heart failure are more susceptible to developing lethal arrhythmias. So, not only does this abnormal myocyte shape lead to excessive wall stress, which impairs blood flow in the heart, it also promotes lethal arrhythmias. We have known for a long time that hyperthyroidism may lead to increased arrhythmias, but our lab showed that hypothyroidism also promotes arrhythmias. Additionally, we showed that rats with elongated myocytes from MI are also more susceptible to induced arrhythmias. Myocyte shape matters!

The **SHHF** (Spontaneously Hypertensive Heart Failure) rat model caught my attention. Normally experimental rats can live up to about 3 years. SHHF rats began to die of dilated heart failure at about 18-20 months of age. No surgery is required since this is a genetic disorder. Myocardial tissue is largely composed of contracting myocytes (no MI) with an increase in intervening connective tissue at later timepoints being significant in SHHF rats, bult relatively minimal. At the time of heart failure development in SHHF rats, heart weight is double normal, providing a very large window to detect even modest improvements with drug treatment.

We were among the first labs in the world to acquire the newly developed class of drugs known as AT-1 blockers. This class of drugs targets the cell membrane receptor, the AT-1 receptor, that Angiotensin 2 binds to in arteriolar smooth muscle cells in addition to cardiac myocytes. Activation of the AT-1 receptor triggers contraction of arterioles and raises blood pressure. ACE inhibitors target an enzyme involved in converting Angiotensin 1 to Angiotensin 2, which occurs in the lungs. Angiotensin 2 also stimulates the release of Aldosterone in the kidneys, which inhibits urine output, leading to increased blood volume. It's easy to see how overactivation of Angiotensin 2 (increases blood pressure) and Aldosterone (increases blood volume) in heart failure can overload and stress the heart. So, reducing blood pressure and blood volume are logical targets for these patients. ACE inhibitors and AT-1 receptor blockers serve the same purpose, to reduce activity of Angiotensin 2. Some patients taking ACE inhibitors develop a dry cough, a symptom that is minimal with AT-1 blockers. So, AT-1 blockers were an important advance in treatment especially for these patients.

In our experiment, SHHF rats were treated with an ACE inhibitor and an AT-1 blocker from Merck. Additionally, we treated another group of rats with the vessel relaxer, hydralazine. The Pfeffer's had previously shown that hydralazine reduced blood pressure in hypertensive rats but not the resulting heart enlargement or improvement in survival. Since Merck's AT-1 blocker, Losartan, was already providing good results in patients, they gave us another AT-1 blocker in their arsenal in case our results were not promising. Merck chose Losartan because they had a global patent but only the US patent for the AT-1 blocker we tested. The resulting study published in 2000 and featured on the cover of *Circulation* was remarkable **(Figure 11)**(20). We treated cohorts of SHHF rats with these 3 drugs from 18 to 22 months of age, as they were progressing to end stage heart failure. At this late stage, left ventricles of SHHF rats are extremely large and dilated and myocytes have an increase in diameter (early change from hypertension) and length (late change associated with dilation). Both the ACE inhibitor and the AT-1 blocker reversed remodeled myocytes literally back to normal. In this case, SHHF rats clearly have a hyperactivated RAAS, which is specifically targeted by ACE inhibitors and AT1-blockers. The study also showed for the first time that elongated myocytes in terminal heart failure had the ability to remove sarcomeres in series while shrinking the dilated ventricle back toward normal. To this day, we have no idea how this occurs. **This was the first study ever to show that end stage dilated heart failure is literally curable pharmacologically at the cellular level if the drug matches the underlying cause**. In the other parallel group, hydralazine had normalized hypertension in SHHF but had no effect on heart weight or myocyte size. This confirmed the Pfeffer's observation but with additional myocyte size data. At the next AHA meeting, I noticed that Merck was passing out copies of this Circulation paper like candy. I later confirmed that their drug marketed in humans, Losartan, was equally effective in this rat model.

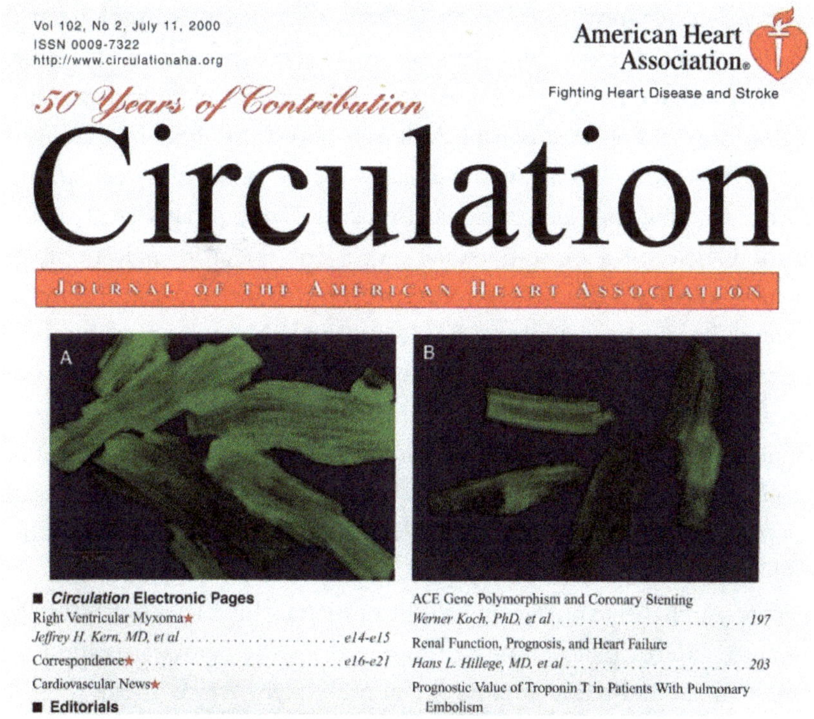

Figure 11. Reverse remodeling of myocytes in SHHF rats. Severely enlarged isolated myocytes on the left were from untreated SHHF rats in failure. Myocytes on the right were from SHHF rats treated with an AT-1 blocker and were normal in size. Contractile proteins, labeled in green, show that these proteins remain intact even in heart failure. The myocyte shape change was important in recovery.

The Chairman of Medicine at Harvard, Victor Dzau, invited me and another researcher to give the keynote dinner addresses at the annual Merck consultant meeting the following year. These were exciting times in my career. I became close friends with many of the world's leading cardiologists and heart researchers. Quite a few chiefs of cardiology at top medical schools tried to recruit me. But, in the end, I chose to take the anatomy department chair at the University of South Dakota. I figured that if I screwed up, nobody would notice being out in the middle of nowhere. Thankfully that did not happen. At age 40, I was the youngest anatomy department chair in the US. Of course, as I left Tampa, the Chairman of Medicine at the University of South Florida (USF) told me that I would never be heard from again. I had heard that before when I left UAB for USF and was intent on proving him wrong.

Shortly after moving to South Dakota, the chief of cardiology at the University of Colorado, Mike Bristow, invited me to Denver. He brought me there to give a private seminar on cardiac remodeling to Eric Olsen. At that time, there were considerable rumblings that Eric might win the Nobel Prize for his discoveries related to the molecular regulation of growth and development of skeletal muscle. During that visit, I first got to know Eric on a more personal level. Eric is surprisingly humble, a down-to-earth guy who is always respectful to everyone. Before my talk, he said: "I am excited. I know I will learn a lot." No pressure. You should know that Eric is probably the most gifted scientific speaker I have ever heard. His talks typically draw the largest audience at any meeting. He connects with everyone. You could ask someone leaving his talk to summarize key points and literally anyone could do it even if their research was in another field. Shortly before this encounter, one of Eric's postdocs published what would arguably become the most influential paper ever in heart research. So, Eric decided to shift much of his focus to heart failure research, an area that had more funding from the National Institutes of Health. Eric had made a name for himself at the Texas Medical Center in Houston. As his fame grew, the University of Texas Southwestern in Dallas recruited him. I heard someone joke that they had to mortgage the state of Texas to recruit him. He has a huge operation in Dallas. It is still mind-boggling to me that someone as busy as Eric nearly always returns my emails within minutes or hours. He once visited us in South Dakota. Surprisingly, I learned that his dad attended Augustana College in Sioux Falls and got a master's in chemistry at the University of South Dakota in Vermillion. Eric was frequently on his cell phone reminiscing with his dad during the visit. Eric was born after his parents left South Dakota. I sometimes joke that if you were to ask 1,000 heart researchers, who is the best in the world? About 900 would name Eric and the other 100 would be envious as hell.

Eric and Mike Bristow started a biotech company named Myogen. They would later sell it for $2.5 Billion. Mike gave me some stock in the company after I submitted a patent application through his company for a new technique I developed. Unfortunately, I let my wife talk me into cashing out for $10,000 to purchase antiques just before the company was sold and the stocks took off. Good thing I didn't go into investment as a career.

I would like to end this chapter with a brief discussion of another type of heart failure. During my early career, dilated heart failure *was* heart failure. It is now known as **HFrEF** (heart failure with reduced ejection fraction). With end stage HFrEF, ejection fraction is reduced from ~70% to less than 20% near death. Ejection fraction is the percentage of blood when maximally filled that gets pumped out of the left ventricle with each beat. Mike Zile, from Charleston, started telling everyone that patients with hypertension were dying of heart failure but didn't have dilated hearts. There was later a humous presentation at a meeting of the Heart Failure Society of America by another well-known German cardiologist entitled: "Why is it that patients with heart failure, don't have heart failure?" He was referring to the newly discovered non-dilated form of heart failure. This became known as **HFpEF** (heart failure with preserved ejection fraction). My friend Maggie Redfield, from the Mayo Clinic, later published a landmark paper in the New England Journal of Medicine reporting that about *half* of heart failure patients actually have HFpEF(21). This was a rather shocking wake-up call to heart failure cardiologists. HFpEF patients are typically elderly women with hypertension but also includes others. It quickly became known that standard drug treatment regimens for dilated heart failure were not working for patients with non-dilated heart failure. While patients die of non-dilated heart failure at the same rate as dilated heart failure patients, they are generally excluded from heart transplant programs since reduced ejection fraction is a selection criterion. While the mechanism of HFpEF is still being explored, common features are increased collagen between myocytes and impaired coronary microvascular blood flow (small arterioles rather than large arteries).

I asked my dean at USD, Bob Talley: "How could this have been overlooked for so long?" He told me an interesting story. In his early career as a cardiologist at the VA, he noticed patients with non-dilated hearts were having symptoms of heart failure. In an experiment to better understand this, he overloaded their hearts with an intravenous infusion of saline. They briefly got much worse. He called this phenomenon "Stiff heart" because the left ventricle could not accept sudden increases in blood volume as it would normally. When you consider that collagen has the tensile strength of steel and accumulates between myocytes during hypertension, it is not hard to imagine how more collagen between myocytes can make these hearts stiffer. Bob submitted a paper to the New England Journal of Medicine summarizing his findings. The manuscript was returned with the editor saying: "This is the most ridiculous thing I ever heard of." So, it was there all along but ignored because it did not fit with the understanding of heart failure at that time. These were among the heart failure patients who did not respond well to newer treatments. So, it may have been easier to just ignore them as the unfortunate ones. This is a big need in cardiology, although some recent drugs appear to offer hope.

> *Throughout my career, heart failure meant dilated heart failure. However, a few years ago a non-dilated type of heart failure was discovered (HFpEF). Apparently, it was there all along but did not fit into our understanding of heart failure. So, it was largely ignored until it clearly became the elephant in the room. HFpEF constitutes half of heart failure cases and standard treatments do not seem to work well. Developing new therapies for this disorder is one of the most pressing needs in cardiology. Patients are mostly elderly women with hypertension but others are also included in this group.*

Chapter Eight
Are We Ignoring One of the Biggest Advances Ever in Heart Failure Treatment?

When choosing cardiac remodeling as my research focus, I honestly felt this was a field where my work could eventually lead to a major advance in the care of patients with heart failure *in my lifetime*. As I am near the end of my career, there is a very good chance I was right. The remaining step is translation to humans, something I am obviously encouraging. I will now describe the step-by-step experiments leading up to this promising outcome to my life's work. I want readers, particularly young people considering becoming biomedical scientists, to understand the process and experience what it is like to walk in my shoes. First, I should say that I am an atypical researcher. I really don't know anyone who thinks quite like me. But isn't that what makes science interesting? We are all uniquely different people. I believe in a common-sense approach to research focused on identifying potential avenues for new treatment, confirming efficacy, and rapidly doing the work needed to implement the treatment in patients.

As I began my career, it was clear that reductionist research was where everyone *must* go. **Reductionism is the strategy where it is believed that complex processes can be broken down into their simpler components and how the parts work together to produce the properties of the whole.** Descriptive studies, my favorite, are more difficult to get funded and considered inferior by grant study sections. However, I have always felt that precise descriptions of fundamental truths are needed to provide a strong foundation for more mechanistic studies. Unfortunately, I have found that many descriptive studies are done so poorly that they do not provide that strong foundation to build on. I remember reviewing a grant once in a NIH study section meeting. The other reviewers raved at how amazing and innovative the proposal was, largely from a mechanistic standpoint. But I pointed out that the foundational data were contradictory and of such awful quality that they could not be trusted. The other reviewers did not know how to recognize the difference between solid and weak descriptive data! The reviewers were going to approve moving forward with funding while the investigators were building their research on a garbage dump! This is why I have been so particular about the accuracy of my studies. You need to trust the foundation if the goal is to stand the test of time.

For the most part, scientists are hard-working, dedicated people on a mission. But people are people, and there are exceptions. I have seen enormous resources wasted, patient health jeopardized, years wasted, and young scientists' careers damaged by chasing after lies. Work by Piero Anversa is a very good example of this. There is an excellent review that summarizes his story(22). This may well be the most damaging story of medical research fraud in history. Many smart people jumped on his bandwagon, often saying "But, what if he is right?" Piero once invited me to visit his lab in New York and later took me to a very nice Italian restaurant. Near the end of our visit, he asked me to make a deal with him: "I will positively review your manuscripts and grants if you return the favor". I said: "Of course I will enthusiastically support all good science." He did not like my response. He wanted a rubber stamp. I was the first to question his data publicly and he wanted to win me over. I noticed that his data seemed made up to fit his story and often clearly misleading. Over time, so many jumped onto his bandwagon that he obtained major NIH funding and won the most prestigious research award at the American Heart Association annual meeting. But for those of us who could see through him, he fit the saying: "If you can't dazzle them with brilliance, baffle them with bullshit." Although Harvard may have been influenced to recruit him largely because of the massive amount of NIH funding he had acquired, many heart researchers I knew there considered him a fraud.

It is interesting what finally led to Anversa's downfall and a comprehensive investigation of his work. It was not the cardiac community but the physics community who finally exposed him. Here is the story. Since radioactive carbon is incorporated into dividing cells, many people living near nuclear test sites during the atmospheric testing period after WW2 should have permanently labeled cells that were dividing at the time when they were exposed to radiation. Using this information, another group showed essentially no evidence for cardiac myocyte division in adult hearts. Anversa showed the opposite result in a mirrored experiment. In both cases, key data were processed at the Lawrence Livermore National Laboratory for their respective publications (yes, named after EO Lawrence, the kid from Canton SD). With their reputation on the line, investigators at LLNL discovered serious discrepancies in data sent to Anversa and those reported in his paper. They asked for an investigation. As a result, he was finally barred from NIH funding and is no longer working in the US to my knowledge. Many of his key papers in top journals were retracted. Scientists are not all altar boys. While they may often disagree, uncovering the truth is the motivation for nearly everyone I know involved in science.

Back to an earlier discussion about my approach to doing research. The reductionist approach in my field has largely focused on identification and testing of key molecular signaling pathways implicated in heart failure. Mostly, this is done in mouse models by knocking out or overexpressing a specific gene believed to be involved in the disease process. This process may use a whole-body model or be specific to a given organ, like the heart in my case. Sometimes the specific gene is so necessary that gene-altered animals die. More often, gene targeted mice appear to be healthy. So, researchers that have targeted cardiac specific genes stress the hearts by producing surgical MI or putting a constriction around the aorta to create resistance to blood leaving the heart, resulting in a pressure overload on the heart. If animals with the deactivated gene do better, it is concluded that the gene plays a detrimental role in the disease process. If they do worse, it is concluded the deactivated gene is beneficial or protective. If the gene is overexpressed and after inducing stress, animals do worse, this offers evidence that the gene is a bad player in the disease process. If the overexpressed gene provides benefits, it is surmised that it is a good player and a possible direction for new drug development. Usually, outcomes of these upregulation and downregulation experiments are complementary and support each other. Unfortunately, these experiments have rarely led to effective new treatments in humans. Increasingly, scientists have realized that reductionism is too simplistic an approach to understand a complex, multifactoral disease. Highlighting this naivety is the fact that overexpressing or knocking out a given gene in the heart often leads to changes in hundreds of other genes. So, how do you know that targeting a specific gene is responsible for the outcomes when it could be another altered gene or many altered genes responsible for the observed outcome?

More recent developments are encouraging though. Gene array analysis of animal models typically reveals alterations in key gene families involved in important processes like metabolism or inflammation, for instance. But this still does not get us where we need to be in identifying a specific treatment that works. There are now mountains of human and animal gene array data published for just about every situation imaginable, just sitting there awaiting future studies that can effectively use these data. This is starting to move forward with AI. The two "Holy Grail" anticipated future discoveries rely on the coldest and hottest things imaginable. Quantum computing requires temperatures approaching absolute zero. I would recommend Michio Kaku's book "Quantum Supremacy" to learn more about this exciting field. A controlled nuclear fusion reactor that can run on water without producing dangerous waste requires extremely high temperatures sustained for a given period. Both fields are still a work in progress. However, positive reports are becoming more frequent, suggesting that both fields may soon cross the critical threshold of making these dreams a reality. I am optimistic that both will become feasible within my lifetime.

> ***The near future of medical discovery is exciting.*** *Quantum computing coupled with AI may someday revolutionize research and overcome the reductionist shortcoming of dealing with complexity. For instance, prominent scientists suggest that new treatments for cancer may play out as follows. A scan of all expressed genes (~30,000 genes) is done on a cancer biopsy of the patient. Results will be compared to all relevant data in the medical literature. Specific signaling pathway(s) and key regulatory molecules likely involved in the process are identified. The best drug targeting sites on these regulatory molecules are identified. Protein structure of the most likely molecular shape of the treatment compound is identified. Available drugs meeting those criteria are identified. Especially important are those compounds that have already been tested for safety. If none meet the criteria, all compounds available around the world in the*

inventories of participating drug companies will be searched. If none meets the criteria, a new drug will be designed based on the desired molecular structure. For instance, a drug that should block a key site on a molecule playing an important role in the identified signaling pathway. This information is sent to a specialized company to quickly make the compound. All the computer work will happen in nanoseconds with specific treatment available for that patient within a very short time. It looks like we may be there sooner than you think. Although current computers are far from the power that quantum computers will someday deliver, AI platforms are already making major progress toward the scenario described above.

My quest to find a new treatment for heart failure kept circling back to thyroid hormones, where I began my career. During my PhD thesis work, I learned that hyperthyroidism produced a physiological type of heart enlargement and later showed that the increase in myocyte size was also due to a proportional increase in length and diameter. This is what happens during maturation when heart size increases proportional to body weight to keep up with the increase in cardiac output required by a growing body. The primary driver for heart weight increase is body weight. This is why our heart weight is much larger than a mouse's heart, but heart weight/body weight ratio is relatively consistent between mammals. As mentioned earlier, we would later show that volume overloading from an AV fistula produced a similar myocyte shape pattern as growth-maturation and exercise. But hyperthyroidism is more like non-stop exercise 24/7 and clearly this cannot be sustained indefinitely. We also demonstrated that low thyroid function, hypothyroidism, eventually leads to dilated heart failure with the myocyte increase in length/width ratio resembling dilated heart failure of other etiologies. Of course, having normal thyroid hormone levels is ideal.

Realizing the limitations of a reductionist approach leading to a cure for heart failure, I decided to take a more pragmatic approach that could work in my lifetime. Starting with my belief that cellular remodeling is critical to understanding the cause of heart failure, I developed an "outside in" approach to research. In my case, that meant first proving that an abnormal, maladaptive change in myocyte shape was the underlying basis for increased wall stress in dilated heart failure. I accomplished that goal. Increased wall stress particularly in the resting, relaxation phase of the heart cycle simply had to be a major contributing factor in my estimation. Our hearts beat, non-stop, our whole life. Those brief periods of rest between contractions are critically important. That's when the heart gets perfused with nourishing blood. The next step for me was to identify what makes this abnormal myocyte remodeling process better and what makes it worse. I reasoned this information would put me on the path to an effective treatment.

Funding and publication decisions strongly encourage investigators to focus on the specific underlying molecular mechanism, how it works, with the hope this will lead to identification of new drug targets. But what if your experiments have identified a treatment that already appears to work in animals? Shouldn't you then focus on getting this treatment to patients in need as soon as possible? By that I mean identifying what works is more important than how it works from the patient's perspective. I reasoned that getting a promising new treatment to dying patients as soon as possible should be a top priority. But those reviewing grants want you to focus on identifying the molecular mechanism first, something that could take years or decades to clarify. In the meantime, patients continue to die. I say, why not conduct experiments that accelerate promising new treatments to patients? Then, prop your feet on the desk and design experiments to answer how it works.

Before the development of Synthroid, a man-made synthetic hormone, treatment was with desiccated thyroid hormones containing both T3 and T4 (Armour Thyroid), extracted from pig thyroid glands. Historically, treatment of hypothyroidism is a long and complex story filled with sometimes rather disturbing behavior by drug companies and experts in the field. I would recommend the recent book "Rethinking Hypothyroidism" by my friend Antonio Bianco, who is the world's leading expert on this topic. The medical establishment largely opposes treatment with Armour Thyroid even though a recent meta-analysis of all relevant published clinical studies comparing Armour and Synthroid showed that most patients preferred Armour(23, 24). They simply felt better. In the case of the heart, evidence suggests that in heart failure, the heart may deactivate T4 to inactive rT3, *making itself hypothyroid.* So, there is a strong argument to be made that treatment of heart patients with combined T3 and T4 may be a better choice to ensure that those with impaired ability to convert T4 to T3 obtain effective therapy. Nonetheless, several studies show benefits in heart patients from T4 treatment.

> *It might be helpful if I provided some clinically relevant background on thyroid hormone dysfunction. Although my research expertise is primarily in the area of cardiac remodeling, I have published enough papers in thyroid hormone research to be considered an expert in that area too. If fact, a recent notification indicated that I rank in the top 0.1% of scientists in the world publishing in this area.* **Hyperthyroidism** *is relatively straightforward to diagnose and treat, although it can take a few months to optimize treatment. The most obvious change noted in the patient is a racing heart. Heart rate can exceed 100 beats per minute, while ~70 is normal. Other signs are heat intolerance and weight loss. When diagnosed, the patient is typically treated with an anti-thyroid drug like* ***PTU*** *(Propyl-ThioUracil), possibly thyroid surgery, or radiation ablation of the gland followed by hormone replacement if an anti-thyroid drug does not work. A beta blocker may also be given to quickly reduce heart rate if that presents as an immediate problem.* **Hypothyroidism** *can be more subtle to diagnose. Additionally, a large percentage of patients may be elderly, particularly post-menopausal women and may attribute symptoms to aging. Symptoms of hypothyroidism include fatigue, slow heart rate, weight gain and cold intolerance. Another problem is that patients who have hormone values within the lower end of the normal reference range may be symptomatic for hypothyroidism and can go untreated because their blood values are normal. The drug of choice for treating hypothyroidism is Synthroid, which contains only T4. T4 must be activated in peripheral organs like the liver and kidneys where it is converted to T3 by removing an Iodine (the number indicates the number of Iodine's in the hormone).*

More recently, it has become appreciated that as many as ¼ of hypothyroid patients treated with Synthroid remain symptomatic. This can usually be corrected by adding T3 to their treatment regimen.

While my myocyte remodeling research was ongoing, molecular biology was rapidly becoming the major driving force in heart research. Eugene Morkin(25) and others(26) were studying the myocyte contractile proteins that regulated heart function. They found that in normal rodents, the more rapidly contracting α-myosin was the predominate form, while the slower form, β-myosin, was minimally expressed. It was later found that β-myosin was the predominate myocyte isoform in larger mammals, like humans. This made sense because larger mammals have slower heart rates than smaller rodents. It was also discovered that thyroid hormones serve as a major regulator of myosin isoforms. Hyperthyroidism leads to upregulation of α-myosin and down-regulation of β-myosin while the reverse was true of hypothyroidism. During fetal growth, cardiac myocytes produce mostly β-myosin and thyroid hormones are lower at that time. There is a surge in thyroid hormones after birth, leading to a switch to predominately α-myosin in rodents as they mature. Importantly, it was discovered that pathological heart conditions, such as MI or hypertension, lead to reversal back to the fetal gene expression pattern. The finding that pathological cardiac enlargement was associated with a return to the fetal gene expression pattern captivated heart researchers for many years. But, because human hearts had mostly β-myosin it did not initially seem clear that this isoform shift was important in us. Indeed, for a few years it was not even clear if human myocytes had much α-myosin at all based on protein assessment. I began to wonder if this was irrelevant in humans, why was everyone encouraged to study it? Later, Mike Bristow's group showed that there is indeed a small but significant fetal gene shift in humans too when examining mRNA rather than proteins(27). Overall, activity in this field gave us a very important early understanding of how genes are regulated in the heart. It provided a good foundation as the field of molecular biology matured.

While writing this book, I was saddened to hear that Mike Bristow passed unexpectedly. I saw Mike at a recent meeting. He asked me to go have a beer with him so he could get my input on a perplexing issue in cardiac physiology. He was having a discussion with some other heart failure experts about specific functional properties of the failing heart, and they could not agree. He was working on a review article and wanted to know my opinion. Months later, I was reading the review and came upon the section he discussed with me. He incorporated my input. It put a smile on my face when I read it. That was the last time I saw Mike, one of the most influential heart failure experts of his time and was once nominated for the Lasker Award (prelude to the Nobel Prize). He will certainly be missed by many.

Earlier in my career, I was confronted with a difficult question. Should I continue focusing on structural remodeling of the heart or switch to molecular biology. It is worth noting that I had never had any formal training in molecular biology. There was a lot of pressure to change directions to this field. While attending a small summer meeting at Big Sky Montana, I went on a walk with my friend Ketty Schwartz from Paris. Ketty was one of the most famous cardiac molecular biologists in the world. She later went on to become the equivalent of the French NIH director. Her advice was that I absolutely must continue what I am doing. She said: "I was providing the most precise and comprehensive information on cardiac remodeling in the world. People trusted my work and it was important. Everyone else was doing molecular biology. I had a unique niche". During that walk, we came upon a moose with her young calf. Ketty wanted to get closer, but I quickly grabbed her by the arm and pulled her back, warning her that the moose was getting ready to charge us. So, we slowly backed away and lived to tell the story. Unfortunately, Ketty passed away a few years ago from breast cancer. I miss this wonderful kind and brilliant woman very much.

I got another opinion from Jeffrey Robbins, another leading cardiac molecular biologist. Jeff did something remarkable in his early career. He developed a method to overexpress a selected gene of interest in mice, specifically in the heart. He then gave away this *a-myosin promoter* to the rest of the world for free. There are now hundreds or possibly thousands of mouse transgenic models that used his technology. It was a revolutionary technical advance in the field. Jeff was a dear friend who did so many favors for me over the years, like serving as an external consultant for the research center I started in Sioux Falls. He grew up near where I currently live in New York and dropped by to give a seminar a couple of times. He had an interesting relationship with his demanding mother who he visited rarely. "Jeffrey, if you're so smart, why don't you have any Nobel Prizes (plural)?" she once asked him. As a side hobby in his younger years, Jeff liked to race cars, until his back would no longer allow it. Unfortunately, he died recently as a pedestrian, hit by a young, inexperienced driver. Earlier, Jeff left the University of Cincinnati to conduct research at Cincinnati Children's Hospital. He is largely responsible for CCH becoming one of the leading children's research hospitals in the US. The advice that Jeff gave me was a bit different than that of Ketty. He told me, "I needed to reinvent myself and incorporate molecular biology into my research." Realizing that the field was stocked to the brim with smart young molecular biologists, I took his advice. I came up with my translational projects with the goal of identifying new treatments for heart failure (in my case thyroid hormones) and recruited these young people to come up with sub-aims in my grants addressing the underlying molecular mechanisms. This worked out well for everyone. Indeed, these smart young people often told me they were attracted to my lab because of the clinical-translational focus, not wanting to be another molecular biologist spending their life learning "more and more about less and less". They saw meaning in what I was doing. I encouraged my best PhD student, XJ Wang, to do a postdoc with Jeff Robbins. Jeff returned the favor by encouraging XJ to rejoin my group as we started to build a new cardiovascular research center in Sioux Falls. XJ is now a world-famous heart failure researcher, linked to both our labs for life. These interconnections between trainees and mentors often become like families. In the end, I took the advice of both Ketty and Jeff. It made a big difference in my career. I believe you now have some understanding of the remarkable friendships one can develop during a research career. Priceless!

Back to science. So, thyroid hormones produce a physiological type of myocyte hypertrophy and myosin expression pattern. Hypothyroidism eventually leads to dilated heart failure with a myocyte shape change and fetal myosin re-expression pattern characteristic of heart failure. This observation encouraged my "outside-in" approach to research. By that, I mean conducting experiments that gradually drill down to basic truths by identifying what takes you in the right direction while excluding less relevant peripheral information. The next logical step was to determine if superimposing thyroid hormones on pathological models of heart failure leads to improvement in heart function and remodeling. Of course, I realized that giving high doses of thyroid hormones to heart failure patients would never be feasible because of the prospect of triggering arrhythmias. These animal experiments were what is known as *proof of concept.*

> *For the first time, we definitively demonstrated that* **below normal thyroid hormones (hypothyroidism)** *in rats* **causes dilated heart failure.** *It was a slow process but a clear outcome.* **Elevated thyroid hormones (hyperthyroidism)** *produced a beneficial increase in heart function and cardiac remodeling, resembling exercise. But this is like non-stop exercise 24/7 and is clearly detrimental in the long term. In a series of experiments, we superimposed a high dose of thyroid hormones on various animal models of heart failure. In each case, heart function and cardiac remodeling improved. However, these were short-term, proof of concept experiments since overdosing heart failure patients with thyroid hormones was definitely not a treatment option. So, the next step was to treat heart failure models with a therapeutic dose of T3. Realizing this may not affect heart weight or myocyte remodeling, I began to doubt my original theory. But the experiments needed to be done and led to very exciting results offering the final proof for my original hypothesis.*

In earlier experiments, SHHF rats were treated with high doses of thyroid hormones, high enough to produce an additional increase in heart weight. Could this physiological stimulus override the pathological process? These experiments produced very interesting results. High dose treatment did not make the hearts more dilated, which would have been detrimental. Rather, it changed myocyte cross-sectional shape in a manner that normalized the elevated wall stress from hypertension. The myocyte shape change led to a further increase in wall thickness and a reduction in dilatation, changes that reduce wall stress. The rats still had hypertension, and the heart was beating faster, but elevated contractile wall stress was attenuated and heart function improved. It was as if the myocytes knew how to grow in the only manner that was beneficial when faced with a thyroid hormone growth stimulus. These experiments were followed by high doses of thyroid hormones in rats with MI. Again, myocyte cross-sectional shape grew in a manner that reduced contractile wall stress, and heart function improved despite a tendency for heart rate to increase. The progression of these experiments was discussed in a recent review(28).

After the encouraging results described above, it was clear that the next step was to investigate the effects of *therapeutic* doses of thyroid hormones in our animal models of heart failure. People take oral thyroid medications daily. You can't teach a rat to do this. An easy treatment approach had not been worked out in small animals, particularly if one wanted to do long-term treatments. The active form of the hormone, T3, has a half-life of only one day, so daily, or twice daily treatment is preferred. For short term experiments, many rodent studies used daily intraperitoneal injections in the abdomen. But it is a bit painful stabbing them in the belly daily, not feasible for long term treatment, and introducing stress into the experimental results. Scientists are, of course, required to minimize potential pain to research subjects and we take that mandate seriously. Realize that results from animal studies also benefit our pets with new treatments too. I will give an example of that later. We tried mixing thyroid hormone powder into food pellets. But some rats started to eat more, increasing the dose. Since adult rats are typically housed two per cage, one rat may also be a more aggressive eater, producing more experimental variability. So, food restriction was a necessary complication. We also tried implanting subcutaneous slow-release pellets containing thyroid hormones. But the supplier sometimes got the dosage wrong, literally sabotaging our experiment on two occasions. Some approaches also made it necessary for research technicians to come in on weekends since animal facility caretakers who change water, food, and bedding do not generally get involved in specialized treatments. I realized that it would be ideal if T3 could be added to the drinking water. This would make it easier to adjust doses if necessary and the animal support staff could implement the protocol by simply using treated water that we provided. But thyroid hormones are not water soluble. That problem was solved by dissolving T3 in an alcohol solvent, then mixing the desired dose in water based on the animal weight and normal daily water consumption. This method was not stressful to animals, dosage could be easily adjusted, and long-term studies could be easily conducted after establishing optimum dosing. This became our treatment protocol of choice using the active form of the hormone, T3. We have now published many studies using this approach. The initial studies examined low, moderate, and high doses to establish the "Goldilocks" dose, not too high, not too low, just right. Since excess T3 stimulates heart rate increase, a useful target was to determine the dose that increased heart rate, then back off a bit so heart rate normalized. Realizing that the optimum dose would not affect heart mass or heart rate, it was not clear what to expect. But the experiments needed to be done to confirm potential benefits that could advance clinical trials. Admittedly, I was a bit apprehensive that these experiments could undermine my goal of using thyroid hormones as a valuable treatment of heart failure. It was completely possible that low dose thyroid hormones, the only treatment option, would not produce detectible improvements.

In rats with MI, therapeutic doses of T3 did not affect heart mass, as expected. But heart function was clearly improved and there was a reduction in the size of the infarct with better preservation of viable myocardium at the boundary of the infarction(29). SHHF rats(30) and rats with diabetic cardiomyopathy(31) also showed improved heart function. T3 doses were too low to induce further enlargement of the heart. But why were myocytes performing better? Was there beneficial remodeling inside the cells? More about that later.

Importantly, our lab and that of Warner Simonides in Amsterdam discovered that pathological heart diseases trigger a reduction of thyroid hormones in the heart, specifically a reduction in the active form T3. Unfortunately, there was only one lab in the world measuring tissue levels of thyroid hormones using **MS** (Mass Spectrometry), the most accurate method, that of Riccardo Zucchi and Alessandro Saba in Pisa, Italy. So, we began a collaboration with several visits back and forth. Who doesn't want to visit Tuscany? We also started a collaboration with Giorgio Iervasi and Alessandro Pingitore (also from Pisa) who have conducted more clinical trials with heart failure patients treated with thyroid hormones than anyone in the world. Currently, however, more long-term clinical trials investigating the efficacy of therapeutic thyroid hormone treatment of patients with heart failure are needed. Importantly, there is not an established protocol in clinical practice for identification and treatment of heart patients with borderline low thyroid function who will likely benefit.

While conducting these animal experiments, many clinical studies continued to suggest that low thyroid hormone function in patients with heart failure leads to poor outcomes and increased mortality. Recent studies even demonstrated that heart failure patients with serum thyroid levels in the lower portion of the *normal* reference range lead to worse outcomes(32, 33). These patients with low *normal* thyroid levels are not being treated and current guidelines would be strongly opposed to it. Furthermore, treatment is typically avoided in patients with borderline low thyroid diagnoses like **Subclinical Hypothyroidism** (normal T3 and T4, but elevated TSH) or **Low T3 Syndrome** (normal T4, normal TSH, but low T3) where the decision is more discretionary. Note: TSH is very useful in diagnosing thyroid dysfunction because it regulates serum levels of T3 and T4. Clinicians are careful with administration of thyroid hormones because heart failure patients already tend to develop more arrhythmias and excess thyroid hormones are known to trigger arrhythmias. Without further understanding, the hesitation to treat with thyroid hormones is understandable. More guidance is needed. Of relevance, we have also observed that hearts from hypothyroid rats are more susceptive to developing arrhythmias and human studies support this. Apparently, thyroid hormone levels need to be in the "Goldilocks" zone to minimize arrhythmias.

An important observation from our lab and that of Dr Simonides is that every animal model of heart failure examined thus far demonstrated *below normal* cardiac tissue levels of the active hormone, T3. This includes SHHF rats, rats and mice with MI, rats with diabetic cardiomyopathy, and hypothyroid rats (as expected)(30, 31, 34, 35). Of note, the diabetic rats had normal serum thyroid hormones but cardiac tissue T3 was below normal. All these studies were done with the Pisa group using MS analysis of T3, a method validated for accuracy. Using another assay method, two studies done in humans with heart failure also reported low cardiac tissue levels of T3(36, 37). But we were unable to validate the accuracy of their method.

Cardiac tissue levels of T3 have been examined using a reliable method in many rodent models of heart disease, including hypertension, MI, diabetic cardiomyopathy, and of course hypothyroidism. In all cases, cardiac tissue T3 levels were below normal even if serum T3 hormone levels were normal. Therapeutic T3 treatment restored cardiac tissue T3 levels and improved heart function in each animal model. Many human studies suggest a similar situation but there are currently no available human cardiac tissue data for T3 using a reliable technique (Mass Spectroscopy). Until recently, there was only one lab in the world doing these measurements, our collaborators in Pisa. We have since transferred the method to our lab and hope to collect human data in the near future.

How can we identify heart failure patients with low cardiac tissue T3 levels who may benefit from treatment? It seems apparent from animal and human data that there may be a disconnect between cardiac tissue levels of T3 and serum hormone levels in individuals with heart diseases. It is likely that serum levels underestimate T3 levels in their hearts. For instance, in our rat diabetic cardiomyopathy model, serum thyroid levels were normal, but T3 levels were below normal in the heart. These diabetic rats had severe cardiac dysfunction. Therapeutic T3 treatment restored cardiac function, restored myocardial tissue T3 levels, and reversed fetal gene expression. They still had diabetes but there was no longer any evidence of heart disease! Another interesting finding was that pathological changes in small arterioles, suggesting microvascular blood flow impairment, were reversed with treatment. This is very interesting considering that new data from patients with diabetic cardiomyopathy suggest they have impaired cardiac microvascular blood flow(39). Could this be corrected with T3 treatment in these patients? I have brought this up twice in sessions at national meetings where presenters were lamenting that cardiac microvascular blood flow impairment is likely a big problem in diabetic patients. They just looked at me like I was from another planet since this had never crossed their mind. Hopefully, I have triggered some interest for them to investigate this.

There are other important questions. How do we know how much T3 or T4 is needed? How do we fine tune dosing without accidentally inducing hyperthyroidism? Another problem is that patients are typically treated with T4 alone. But, for some patients, T4 may not be adequately converted to active T3, which occurs primarily in peripheral tissues like liver and kidneys, but also in the heart. Indeed, about 1 out of 5 patients treated with T4 (Synthroid) still complain about persistent symptoms(40). Another critical need is to identify a serum biomarker that reflects cardiac tissue T3 levels and is responsive to treatment. I struggled with this problem for a while until realizing that we may already have this important biomarker, and it is one of the most widely used biomarkers in cardiology!

As indicated earlier, hypothyroidism induces re-expression of cardiac genes normally expressed before birth. There are other important cardiac genes besides the myosin isoforms that are associated with fetal to adult transformation. **SERCA** (Sarcoplasmic/Endoplasmic Reticulum Calcium ATPase), a gene that regulates uptake of calcium during myocyte contraction, is downregulated in hypothyroidism and heart failure. It is positively regulated by thyroid hormones. One of the most common cardiac biomarkers, **BNP** (Brain Natriuretic Peptide) has been very useful in diagnosing the severity of heart failure. The higher the serum BNP levels are, the worse the heart failure(32, 33, 41-45). From the literature, I also noticed that BNP is high during fetal growth and declines shortly after birth as thyroid hormones surge(46). We know that BNP is produced by cardiac myocytes in the heart when the tissue is stressed. For instance, excessive stretching of myocytes when blood volume load increases, as in dilated heart failure, increases BNP. Of importance here, I noticed an excellent inverse correlation in clinical studies between BNP levels and serum T3. The lower the T3 levels, the higher the BNP levels and vice-versa. Could BNP be negatively regulated by T3? If so, it could be an ideal serum biomarker reflecting cardiac tissue T3 levels. Why would it be ideal? We need a serum biomarker produced by the heart and secreted into the blood when T3 levels in myocytes are low (negatively regulated). Conversely, restoration of cardiac T3 in myocytes with treatment should lower myocyte BNP production and, thus lower serum BNP levels. This should be helpful in guiding treatment in combination with thyroid hormone assays from blood samples.

We designed a series of rat experiments to confirm our suspicion. First, we induced hypothyroidism in rats using an anti-thyroid drug (PTU). After 8 weeks, rats were severely hypothyroid. This protocol causes severe depression of heart function, a reduction in cardiac tissue T3 levels, and induction of the fetal gene program. We treated a parallel cohort of hypothyroid rats with a restorative oral dose of T3 and examined results at 3, 6, and 14 days. Hypothyroid rats had an elevation in serum BNP and increased expression of BNP **mRNA** (messenger RNA) in their hearts. mRNAs provide the message from genes to produce proteins in the ribosomes. T3 treatment quickly led to improved heart function, reversal of fetal gene expression, and a rapid reduction in serum BNP and cardiac tissue BNP mRNA. To simulate a clinical scenario, MI was surgically induced in rats who were divided into T3 treated and untreated groups. This model investigated the potential of administering T3 at the time of MI to determine if treatment slows progression of pathophysiology. After 8 weeks, data were collected. Rats with MI only, had an elevation in serum BNP, cardiac tissue BNP mRNA, and re-expression of fetal genes. T3 treatment reversed each of these changes while improving heart function(47). These exciting results confirmed our suspicion that BNP should be a useful biomarker to detect and treat low T3 conditions in cardiac tissue. Importantly in diagnosis and treatment, it gives clinicians a serum test to confirm if T3 is getting into the heart and normalizing tissue T3 levels. So, they don't have to guess how much T3 is needed.

Many clinical studies also suggested that BNP is negatively regulated by T3. Studies show that low *normal* serum T3 is associated with increased mortality and worsening of cardiovascular function in heart failure patients. Additionally, heart failure patients have a very high incidence of hypothyroidism and borderline hypothyroidism. Recent studies also show that the ratio of T3 to T4 can provide useful information(32, 48). When this ratio declines (less T3 relative to T4), outcomes are worse for heart failure patients. This likely indicates a problem with converting T4 to active T3 in peripheral tissues like the heart. The thyroid gland produces mostly T4 which is predominately converted to active T3 peripherally. So, we can use serum T3, T3/T4 ratio, and BNP to fine tune treatment. We have never been better prepared to conduct clinical trials. What is also amazing is that we are not talking about a man-made drug with a plethora of side effects. We are talking about a hormone that is produced in the body and normalization *eliminates* effects of the hormone imbalance. There is really nothing like this in the cardiologist's arsenal.

An important roadblock to translation of our animal work to humans continued to bother me. How would the physician know the amount of T3 needed to correct the deficiency in the heart? Would you have to push patients to the brink of hyperthyroidism, which could be too dangerous? What was needed was a biomarker secreted by myocytes into the blood when cellular T3 levels were below normal. This serum biomarker should also be quickly reduced when T3 treatment restored myocyte T3 levels to normal (negatively regulated). It was one of those OMG moments when I realized we might already have one and it was already the most widely used biomarker in heart failure diagnosis! Serum ***BNP*** *(brain natriuretic peptide) levels rise as heart failure worsens and tends to decline with effective treatments. A close examination of the literature revealed that this could be the needed biomarker to manage T3 treatment in patients. Many clinical studies consistently showed an inverse relationship between serum T3 and BNP levels in heart failure. In a series of rat experiments, we confirmed BNP was negatively regulated by T3, and T3 treatment of rats with MI confirmed its efficacy for this purpose.*

Readers may be shocked to find that we already knew the potential of thyroid hormone treatment of heart failure patients 75 years ago. In 1950, WB Kountz (49) showed a mortality reduction in 3 groups of men with prior MI who were treated with Armour thyroid (T3+T4). Rather than a proper follow up to this study, patients were overdosed with thyroid hormones in a later study, and researchers reached the conclusion that this treatment was too dangerous. An additional "problem" is that drug companies are not likely to benefit financially. Thankfully, the tide is starting to turn, and the NIH is pursuing a study of T3 treatment of patients with non-dilated heart failure (HFpEF as discussed earlier). This is a condition in serious need of a new treatment since currently available drugs are not very effective.

Another interesting development in cardiology is the realization that patients with open, unclogged coronary arteries may also have inadequate cardiac blood flow. It's not just patients with coronary artery disease. This condition known as impaired **microvascular blood flow (MBF)** is due to inadequate relaxation of small coronary arterioles supplying the heart. MBF impairment has been previously described in patients with dilated cardiomyopathy and hypothyroidism, which was reversed with hormone treatment. Until recently, MBF measurements were impractical for patient diagnosis due to technical limitations. We are rapidly moving to that point, but it is not yet a common part of diagnostic cardiology. Again, thyroid hormones are a powerful regulator of MBF.

We have entered the age of personalized medicine and thyroid hormone treatment of heart failure fits that scenario well. A little background on clinical trials may be helpful. Clinical trials that gave us major blockbuster drugs like Beta blockers and ACE inhibitors often have a fundamental weakness. After a target drug dose is determined, patients are typically treated to reach that dose. While there may be a very significant overall outcome, there is often a subgroup of patients who show no benefits. After treatment is approved and implemented in the general population at this recommended dose, there will likely be a significant number of patients who show minimal or no benefits. But treatment often continues with these drugs even though they may not offer significant benefits. Heart failure patients typically take 7 different drugs per day and put up with all the associated side effects of these drugs. I wonder how a perfectly healthy person placed on this regimen would feel. Additionally, new drugs must show efficacy on top of drugs being taken as standard therapy. It may be that the new drug is as good as the combined effects of other drugs already being taken. But, how would you know? In football, they call that piling-on. The beauty of T3 treatment is that it fits perfectly with the concept of personalized medicine since it can be fine-tuned for each individual with literally no side effects when optimized. You would simply be restoring normal physiological function.

> ***When you consider the likelihood that most patients with heart failure likely have below normal cardiac tissue T3 levels and impaired MBF, especially those in later stages of heart failure, this treatment could be a breakthrough. Additionally, T3 treatment should also improve cardiac contraction and relaxation while reducing fibrosis and atherosclerosis.*** *I believe you may now understand what I was talking about in the introduction regarding optimism about reaching my lifelong goal as a scientist.*

I have found that basic truths can be ignored for only so long. Overwhelming information suggests that most patients with heart failure are likely to have a cardiac thyroid hormone deficiency, a condition that can cause heart failure by itself, superimposed on their diagnosed condition (MI, hypertension, diabetic cardiomyopathy, etc.). We should identify those patients and treat them, and we now have the necessary diagnostic tools. I do not think that thyroid hormones are a "cure" for heart diseases leading to failure. But low cardiac T3 is very likely a major accelerator of progression and overall malaise. After all, thyroid hormones regulate contraction, relaxation, and coronary blood flow while inhibiting fibrosis and atherosclerosis. Heart failure patients also fare worse when depression arises. Low T3 is also linked to depression. I predict this will eventually become a standard part of cardiology treatment and will likely lead to improved quality of life and delayed progression to end stage heart failure. There will always be a need for other cardiovascular drugs since T3 will not cure hypertension, coronary artery disease, etc. I don't think drug companies should be so fearful of this.

This is also personal to me. My dad had a MI when he was about 84. He was very active and had just completed a colonoscopy when this occurred. Before leaving the clinic, he started getting chest pain and was eventually diagnosed with a massive heart attack. They were surprised because he had low cholesterol levels and was active and fit. By the time he was nearing 90, they essentially sent him home to die, saying there was nothing else they could do for him. His cardiologist was not receptive to thyroid hormones because his serum T4 value was low but still within the normal range. I asked my sister Delores, who was taking a low dose of **Synthroid** *(man-made form of T4), to give him 5 pills and asked him to take 1 per day. I knew this dosage would not be risky since his thyroid levels were borderline low and it should only raise blood levels to the midpoint of normal if that. Bottom line: they sent him home to die after exhausting "all" treatments. I could not bear living with myself if this was not tried. A few hours after taking the first pill, he was out of bed and feeling much better. The next day a friend was over playing dominoes while they shared a couple of beers. He could not believe how much better he felt. Still, with this information the local doctor would not initiate treatment because his blood thyroid levels were "normal." Thankfully, I am well connected in the world of heart failure. I contacted my friend Heinrich Taegtmeyer, a heart failure cardiologist at the Texas Medical Center in Houston, not far from my dad. Henrich is probably the world's leading expert on cardiac metabolism in addition to having a heart failure clinic several times a month. He agreed to see my dad. Heinrich carefully mapped out regional heart function in my dad and found that only the inter-ventricular septum (between left and right ventricles) showed good contractile activity. His ejection fraction was very low at 20% (normal is 50-70%), low enough to be on a transplant list if he was younger. Dad showed a remarkable improvement with T4 treatment despite knowing that treatment could not resurrect dead myocytes. Surprisingly, other parts of the left ventricular free wall showed some contractile improvement. This was likely from "reawakening" of patches of myocytes imbedded in scar tissue after the MI. My dad would sometimes cry when he asked me: "Why can't other patients get this treatment? Do you need to have a son who is a heart failure researcher?" Dad lived 2 more years after initiating treatment until he had another heart attack and passed away at 92. It* *was clear that his final days were of higher quality than other patients in his condition.*

I will also share another episode involving my oldest daughter, Meredith, and her cat, Fiona. Fiona had been low energy for several years. She had an acute lung infection and the vet put her on an antibiotic but also pointed out that she may have heart failure, requiring further testing. I insisted that the vet measured T3, T4, and TSH levels based on other results from Fiona suggesting hypothyroidism. Apparently, vets do not typically measure T3 in cats, but she agreed to put in a special order. Fiona's T4 and TSH values were borderline hypothyroid but within the normal reference range. T3 was undetectable. T3, the active form of the hormone was undetectable! Notably, Fiona had lost 30% of her body weight and her heart rate was half normal. I pointed out that these were textbook signs of hypothyroidism, but the vet refused to treat her since it was standard protocol to go by T4 and TSH levels only. As we watched Fiona slowly dying before our eyes, I asked Meredith to give her a tiny piece of an Armour thyroid pill. I am hypothyroid and take Armour. Over the next 2 hours we literally saw Fiona arise from the dead. Later that evening, she was playing with the dog and other cats for the first time in years. Nonetheless, the vet refused to treat her. Based on her body weight and my body weight-dosage, Meredith began treatment that I continue to provide. One pill is enough for about 10 days. Two years later, Fiona had been leading a normal life and remained playful and active.

Unfortunately, Meredith thought she was cured and discontinued treatment. A few weeks later, she rushed her to a different vet clinic. Again, I requested the vet measure T3, which was undetectable again (TSH and T4 were within normal range). Unbelievably, this vet told me that T4 is the active form of thyroid hormone in cats and T3 is irrelevant. She refused to treat her and hung up on me when I tried to correct her misunderstanding of thyroid physiology. I was shocked! I did not think it was possible for anyone to graduate from medical, veterinarian, or dental school with this level of misinformation. Meredith rushed Fiona home and gave her the medication. She quickly recovered from a near death *experience and is back to her normal self. Apparently, the veterinarian field is even further behind than the medical field on this topic. In Fiona's case, the vet's response in both cases was that she had heart failure and immediately needed to see a veterinary cardiologist. The bill could have quickly reached over $1,000 and certainly would not have cured her hypothyroidism. A search of the literature reports that hypothyroidism is extremely rare in cats. However, one report indicated that it is more common than believed. But that report never even mentioned T3. Maybe that is because they don't measure the active form of the hormone, T3. We had another Persian cat, Lily, who died of identical symptoms about 14 years ago. Clearly, more work needs to be done in cats to clarify the incidence of hypothyroidism and its proper treatment.*

You should know that restoration of normal thyroid function in non-cardiac patients is an area of major dissatisfaction for many, mostly women, who have a higher incidence of hypothyroidism. Their chief complaint is that they are still symptomatic despite having "normal" T4 levels. The literature suggests that as many as 1 out of 5 patients treated with Synthroid (T4 only) remain symptomatic. It appears that this issue is a general problem in endocrinology and likely due to the inability of some patients to convert enough T4 to active T3 peripherally. Surprisingly, T3 may not even be included in the thyroid panel ordered by clinicians. I like to compare TSH to the room thermostat setting and T3 to the actual room temperature. High TSH indicates hypothyroidism and low TSH indicates hyperthyroidism. But the thermostat may be broken. It's like the thermostat is set at 70 degrees but you are freezing your ass off and actual room temp is much lower. I would recommend asking for T3 values if thyroid dysfunction is suspected.

When I moved to New York to become chairman of the Biomedical Sciences Department at NYIT, I managed to recruit Dr Kaie Ojamaa, who worked at a nearby research center. I had never met Kaie but considered her the world's leading expert on thyroid hormone molecular signaling in the heart based on her publications. In other words, she was a perfect collaborator for me. Based on the name, I expected to meet a Japanese man or woman. But she turned out to be a blond hair, blue eyed woman whose family was from Estonia. Kaie is an avid skier and biker with a great personality. We both believe in eating well and staying fit. We have become close friends and published many papers together in recent years. Fortunately, we acquired a nice array of high-tech microscopes, including a nanoimager. This microscope can visualize interaction between molecules. This enabled us to provide the final bookend to our animal research regarding setting the stage for clinical trials.

As I mentioned earlier, therapeutic doses of thyroid hormones did not alter heart weight in our animal models but improved heart function. What was going on inside the myocytes? A few investigators had reported pathological changes inside myocytes in patients and animal models of heart failure. Primarily, they observed adverse changes in structures that regulate calcium handling and contraction/relaxation of cardiac myocytes. These structures that regulate myocyte contraction and relaxation, T-Tubules, sarcoplasmic reticulum (**SR**), and SERCA (mentioned earlier) are involved in movement of calcium within myocytes. I knew that thyroid hormones had been shown to affect these structures. In a project led by Amanda Charest, she and Dr. Ojamaa confirmed malformations in T-Tubules and SR in hypothyroid mice like those reported by others in heart failure, including humans. These changes were reversed with T3 treatment. Rats with MI also had maladaptive alterations in T-Tubules and SR. Co-treatment of MI rats with T3 treatment prevented these changes(50, 51). So, my "outside-in" approach eventually led us to specific intracellular changes in myocytes that are responsible for contractile dysfunction in heart failure that are due to low intracellular T3. It's time to focus on translating this to patients. Why are we dragging our feet? I have seen this conservative scenario play out before in my long career.

Not so long ago, beta-blockers were contraindicated for heart failure. This viewpoint made sense because heart contraction was weaker in heart failure and beta blockers inhibit heart function. But my friend Mike Bristow and others kept arguing for several years that the evidence suggested heart failure patients could benefit from this class of drugs. Eventually, they convinced the field that a clinical trial was needed, and of course, drug companies were all in. Results proved so dramatic that these drugs are now a standard part of cardiology. Millions of people around the world are living longer as a result. I will keep fighting the battle for thyroid hormones because the data are simply too overwhelming to continue ignoring.

My original hypothesis as a graduate student was that understanding abnormal structural changes in dilated heart failure could provide the key to understanding this disorder and developing new therapies. Indeed, that eventually led me to the low thyroid story of heart dysfunction. But although our animal studies showed that low dose therapeutic T3 did not affect heart mass, it led to improved heart function in heart failure. Could intracellular remodeling of cardiac myocytes play a role? New research by several labs showed adverse changes in key structural components inside cardiac myocytes in animals and humans with heart failure. Could this be related to low T3? Thankfully, we recently acquired a high-tech microscope that could investigate such changes. We discovered that these intracellular myocyte structures, T-Tubules and SR, were also adversely altered in hypothyroid rats and changes were reversed with T3 treatment. We also observed similar adverse T-tubule/SR changed in a rat heart failure model, again prevented with T3 treatment. Apparently, I had a revolutionary idea as a 23-year-old student that proved to be true. Remodeling was the key!

To date, I have published over 140 peer-reviewed papers in medical journals. Some of my discoveries were featured on the cover of leading cardiovascular journals like *Circulation* and *Hypertension*, two journals published by the American Heart Association. I have been the principal investigator on over $30M in grant funding from the National Institutes of Health during my career and have won international awards for my research discoveries. During my career, I have spent time at 8 medical schools. These were top 20, bottom 20, and in the middle regarding prestige, which is largely based on NIH funding. I have had a wonderfully rewarding research career. I also had a ring side seat to observe how successful and less successful medical schools operated. This experience led me into administration.

I have been around so long that I remember smoking sections in airplanes. In my early years in science, there were ashtrays in the chairs at meetings of the American Heart Association. TV commercials often featured doctors promoting their cigarette brand. I remember interviewing at the med school in Winston-Salem and seeing cigarette machines everywhere. After a night in a bar, you could smell last night's clothes across the room. My friend Chuck introduced me to cigarettes my senior year of high school. I managed to quit on New Year's Eve in New Orleans in 1980. But my girlfriend continued to smoke. There was so much secondhand smoke around in those days I wondered if you could even call it "quitting." To psych myself up to quit, I looked at photos of lungs from people who died of cancer. I reminded myself that some girls would not even date a smoker. It was expensive and unhealthy. I grew tired of people saying: "You're a doctor? You do heart research? And you smoke cigarettes?" I quit cold turkey that midnight and have never looked back. I can't believe I ever picked up that habit. But I was certainly not alone. And finally, thank you Mayor Bloomburg for making the bars and restaurants healthier for everyone.

One last thyroid story to end the chapter. Our recent NYIT president, Hank Foley, our first president with a strong research background, dramatically expanded research. At the annual Christmas party a few years ago, he shared with me his frustration about being chronically exhausted. He had low thyroid function and was taking Synthroid. Throughout his career, he was so exhausted that he would often go for a run during the day to try and wake up. He endured many years of misery while his endocrinologist said his thyroid levels were ok. I recommended Dr Bianco's recent book "Rethinking hypothyroidism" and suggested that he might consider adding T3 since he could be one of those individuals who does not convert T4 to T3 very well. Just about every time I ran into him since, the first words out of his mouth are: "You saved my life!" Hank said he felt like he was 30 years younger. In 2023, I was named the "First

University Professor." This was a new award and the highest faculty award ever given at our university. Although I have a respectable list of accomplishments at NYIT, I suspect that Dr Foley's miraculous recovery from chronic hypothyroidism played some role in the award. He brought Dr Bianco here and tried to recruit him. We came close to landing Bianco but did not have the resources to match the University of Texas Medical Branch at Galveston where Bianco is now Dean.

PART THREE

Leadership and Legacy

Chapter Nine
My Observations of Leadership: The Good, The Bad, and The Ugly

There is an old joke that goes something like this. The parts of the body got into an argument over who should be boss. The brain said: "I make all the decisions, so I should be boss." The muscles said: "Without me, you would be an immobile bag of skin and bones. So, we should be boss." The heart said: "I provide nutrition to the entire body. If I stop, you die. So, I should be boss." The GI tract said: "I convert everything we eat into energy to power the body. So, I should be boss." Then, the asshole applied for the job. All the other parts of the body started laughing so hard that the asshole closed. After a while, the brain got foggy, the heart grew weaker, the legs got wobbly, and the GI tract was backed up and in great pain. They all agreed to make the asshole boss, not because they were fond of the asshole, but they knew their life would be miserable unless they did what the asshole demanded. As my career progressed, I began to see some truth in this old joke.

Ok, I have another asshole joke that is too good to pass up. Two drug offenders came before the judge on Friday afternoon. The judge decided to give them a project over the weekend before sentencing on Monday. "See how many people you can get to stop doing drugs." On Monday, the first man said he got 8 people to stop doing drugs. The judge asked him how? "Well, I drew a big circle and a little circle. I pointed to the big circle and said: This is your brain before drugs. Then I pointed to the little circle and said this is your brain after drugs." Ok, good job. The judge then asked the second man about his outcome. The second man said: "I got 869 people to stop doing drugs your honor." The judge said: Wow, how did you do that? He responded: "I drew a big circle and a little circle. I then pointed to the little circle and said: This is your asshole before prison…."

For some reason, I did not pursue a leadership role in my early career. Yes, I became a leader in the boy scouts, but mainly because I was further along in rank and the other boys did not want the job. I never ran for a student office in school. After becoming a scientist, I focused exclusively on my research agenda. I had no intention of getting into administration at that time. And yet, when a graduate school classmate asked me at the beginning of my career: "Where do you think your career will take you?" I said: "I will probably be director of a cardiovascular research institute some day". So, it was in the back of my mind, and indeed, that happened.

As I progressed through the ranks, I began to realize that the wrong type of people often ended up in leadership positions. I not only saw too much mismanagement, laziness, and lack of direction, but I saw bad people making life miserable for others who were working their asses off. I learned that narcissists were especially destructive. There seemed to be too many control freaks who just wanted you to kiss their ass and know they had power over you. Yes, I know, this sounds like any other profession. People will be people.

My first department chair (now deceased) after I became a faculty member had a reputation for laziness and maintaining the status quo. No one in the department had NIH funding aside from me. My chair got into work about 10am, took 2-hour lunch breaks, and was typically the first to leave. He was near the end of his career and apparently took the job in FL to ease into retirement. Nonetheless, he had a reasonably productive research career prior to becoming chair. His approach with faculty was to divide and conquer. Let them fight among themselves so they would ignore him. On the surface, he seemed like a nice guy. It took me a while to figure him out. In a department of about 10 faculty, two of us accounted for about 90% of the departmental research productivity. Another faculty member came to my office and asked me to quit working so hard, implying that I was making him and the others look bad. A leader's work ethic can rub off. Finally, the two of us who were the most productive faculty, departed. This is what bad leadership does to a place. Many of the best people leave. Less productive brown-nosers reinforce the status quo to secure their future.

The school also had an embarrassing dean at that time. He got into trouble when they found he was using school revenue to pay for his wife's membership in an expensive country club. He also said stupid things.

While he was sitting in first class, an onboarding passenger asked him why he was riding in such an expensive seat while being at a state supported med school. His response: "What, me ride back there with *those* people?" I learned about this by reading the Tampa Tribune. The passenger who asked the question was a writer for the newspaper. This dean also tended to punish excellence and reward mediocrity. He targeted one department chair who ran the most productive basic science department based on publications and NIH funding. The dean kept taking space away from that department and giving it to a much less productive department chair who was highly supportive of the dean. This dean was a textbook example of what not to do.

We finally got a dean who was also a world-class researcher. Among the first things he did was ask the chairs for CVs (resumes) of their departmental faculty. My chair was particularly annoyed by this request and later teamed up with other leadership invested in the status quo who collectively had the dean dismissed. When too many of these people are in charge, mediocrity rules. During this time, there were no pay increases for about 6 years. My salary dropped from about the 40th percentile to the 20th percentile as I was trying to start a family. I had funds in my NIH grant for a pay increase but was not allowed to use them. The next dean was well-intentioned and appeared to be doing a good job. They lured him out of retirement to fix the med school, having previously been dean at a top 10 med school. At least he recognized the faculty who were working hard and positively influencing the universities national reputation. When he heard that I accepted the Anatomy chair at The University of South Dakota, he called me up from the Atlanta airport, sick with the flu, and begged me to stay. He said he could match the salary but not provide a chair position. I appreciated this, but told him it was too little, too late. Good deans and chairs know better than to create an environment where the best faculty can't wait to leave. I believe this dean was well-intentioned, but it would take years to repair the damage already done before his arrival.

During my first faculty appointment, which lasted 10 years, I took two 6-month sabbaticals to learn new techniques for my research program (a 1-year sabbatical is typical). The second med school I visited (after

Zimmer's lab in Munich) was poised to enter the prestigious top 20 in NIH funding. I thought it would be similar to my experience at the University of Alabama at Birmingham. Fortunately, UAB has continued to maintain a top 20 ranking for many decades now. It remains a good example of how to build excellence in a difficult place. But shortly before I arrived for my second sabbatical, they recruited a dean who simply could not stop making bad decisions. The dean fired the chair of a basic science department that was ranked #1 in the nation in their discipline based on NIH funding. From what I heard, it was over a personal disagreement. Other decisions literally gutted some of the best departments in the country. For instance, the dean named the most unpopular faculty member in the division of cardiology, the new chief. Many well-known cardiologists left shortly afterwards. This school has never been able to break into the top 20 and has fallen in NIH ranking. Leadership matters.

I was soaking all this in. I saw the incredible energy and excitement of an outstanding med school that was well run. Then, I witnessed the destruction that one man can do to an excellent med school. I experienced widespread malaise in a desirable location with rudderless leadership. Although business and academia have much in common, they are different. A distinctive difference is that business is more about making money and maximizing profits, while academia is more about building institutional prestige with grants and publications, providing young people with the skills they need to thrive, and being productive with the budget you have. Business and academia both have their share of control freaks and know-it-alls who fail to benefit from the collective wisdom around them. However, fundamental differences are likely why recruiting a businessman to run a university or a government generally does no work out well. The message here when choosing a job is to pay attention to the leadership and overall atmosphere and always remain marketable. Most places have a lot of good people in the trenches and provide an exciting environment conducive to personal growth and achievement. A good chair can also create a very positive environment for his/her faculty even in the background of more disruptive outside forces. But the person at the top sets the tone and has the power to make or break a medical school. Of course, any profession you choose will have these kinds of good players and bad players. Recognizing what to do and what not to do is the first lesson in effective leadership. Sometimes that is not as easy as it sounds.

Chapter Ten
Making the Leap to Leadership

After seeing too much suffering from weak leadership, I decided to try my luck in administration and applied for the anatomy chair at the University of South Dakota. At 40, I would become one of the youngest medical school department chairs in the US. As the recruitment process dragged on, it was clear that I was not the first choice. But, after many months the Dean, Bob Talley, invited me for an interview. He was a cardiologist with a strong interest in heart failure. So, we had much in common. I got the job. I later asked Bob if age was a factor. He said: "Not for me. I became chair of medicine at age 40." Of course, when it's your first rodeo, a major concern on both sides is lack of leadership experience. It helped that I was managing budgets from multiple grants and had a large research group at that time, including graduate students, postdocs, other collaborators, undergraduate and medical student researchers. So, my team was about the size of a small department. Bob was a very good dean. I saw him make a few less-than-optimal decisions, but they were largely driven by the local political environment where he chose the lesser of two evils.

This is probably a good place for an observation I made over the years. I am telling this story because a similar version played a major role in my experience in South Dakota. I have found that university politics can get in the way of good decisions when a state has a major university in a small town and a clear need to expand higher education in one of the state's larger cities. I first saw this at UAB, located in Birmingham, the state's largest city. I could see that the University of Alabama in Tuscaloosa was leery of UAB diverting students and state funding away from them. UAB had grown to a sizeable university before it had a basketball team and did not yet have a football team when I was there. I saw the same animosity between the **University of South Florida (USF)** and the **University of Florida (UF)**. USF had grown to the 19th largest university in the US before it finally got a basketball team. And, like the case in Alabama, it did not get a football team until much later. A similar scenario was playing out in South Dakota. Sioux Falls did not have a state university, but **South Dakota State University (SDSU**) and USD were an hour north and south, respectively. They did not want to see a new state sponsored campus in Sioux Falls, the state's largest city with an ever-increasing need for educating non-traditional students. The 2-hour round-trip commute to either USD or SDSU plus classes in between was inconvenient. The prestigious private college in Sioux Falls, Augustana, was too small and exclusive to meet the overall needs. A school from Colorado expressed interest in establishing a campus in Sioux Falls. This scared SDSU and USD into action. They established a partnership campus in Sioux Falls. But they were so anxious about this campus drawing away students and state dollars, the tuition was much higher. The expectation the new campus would be self-funding was largely intended to inhibit growth. Out of state tuition in nearby Minnesota was actually lower than this new "state sponsored" school. Despite this impediment, it became the fastest growing campus in the state. Each of these scenarios had one interesting driving force that was being overlooked. The urban schools were mainly started to meet educational needs of non-traditional students. They were mostly older residents, often married with children, who desired more education (commuter schools). Often their employer even paid the tuition. Traditional students are quite different. They generally want a real college experience with all the bells and whistles including sports, fraternities, etc. Small town universities often provide a good environment for this, and parents appreciate the walkability (kids may not need cars) and lower crime rate that is generally the case. A graduating high school senior will likely prefer the college atmosphere of Tuscaloosa over Birmingham or Gainesville over Tampa. What 18-year-old would not like to be part of the excitement of competing for a national football championship? More recently, it looks like this has been realized with the paranoia subsiding. Leading small-town universities have grown in recent years and have become more selective due to increased applications. Florida and Alabama are very popular destinations for students across the US with out-of-state students paying higher tuition, also improving the bottom line. In addition to football, the UF has also gone on to win national championships in basketball, swimming, golf, and baseball. If you are a sports fan, this could be your school. I would later learn from an interview for a leadership position at UF that revenue from Gatorade (invented by medical school faculty) generates enough money to fund all these sports at a championship level. Unfortunately, that is a rare situation for other schools that must rely on tuition, ever-dwindling state dollars, and philanthropy. Of note, both USD and SDSU now have Division 1 sports, are doing well, and gaining more national recognition. Obviously, the campus in Sioux Falls made little difference.

I have an interesting side story related to UF. I found myself sitting next to Billy Donovan on a flight from Minneapolis to Sioux Falls. I had no idea who he was but quickly started a conversation at the beginning of the flight. He was the new basketball head coach at the University of Florida. He would go on to win two national championships at Florida and is currently head coach of the Chicago Bulls. Billy was going there to recruit Mike Miller, who would have a great career at Florida before being drafted by Orlando and winning NBA Rookie of the Year. Mike Miller, who I would occasionally run into at Sioux Falls sports bars while he was on break from the NBA, was drafted in the first round and was one of the top high school prospects in the US. It was an interesting conversation on the plane with Billy. I was thoroughly impressed. When Billy arrived at Florida as the basketball head coach, football was king. There was little interest in basketball. Billie was clearly intent on changing that. During our conversation, it was clear that I was speaking to a very driven and talented man with big plans. Getting to know people on their way to becoming famous makes following their careers uniquely special. I am an extreme extrovert. I am often amazed by the outcome of starting a conversation with a stranger. For those introverts reading this, you may consider opening up a bit.

I will now move on to the relevance of this background information to my time at USD. The USD School of Medicine basic sciences (first 2 years of med school) are located on the main campus in Vermillion. The clinical years, 3rd and 4th years, largely take place in Sioux Falls. Sioux Falls is often ranked as one of the best US cities for retirement largely due to its livability, affordable housing, low crime, no state income tax, a nice well-kept zoo, world-class performing arts hall (Yo-Yo Ma commented it had some of the best acoustics he ever heard), home of the South Dakota

Symphony, science museum, an amazing city park system, and exceptional health care. If you ever decide to visit, I recommend going during the Sioux Falls Jazz and Music Festival, a wonderful event. Sioux Falls has two top 100 US hospitals in addition to a VA hospital. Of course, it makes more sense to have a consolidated campus with both basic and clinical sciences. This is particularly true in recent years where integration of clinical scenarios into the first two years has been rapidly increasing. It has always been difficult to get clinicians from Sioux Falls to cancel a half day of clinic and drive to Vermillion to lecture. Zoom has made that less of a problem, but teaching clinical skills is best done in person. For instance, students learn how to use a stethoscope and give physical exams during the first year, skills best taught in person. A large percentage of students commute to Vermillion from Sioux Falls during their basic science years. Roads can be icy and treacherous in the winter. Many also have spouses who work in Sioux Falls, where there are far more job opportunities.

The Basic Science building in Vermillion was in horrible condition when I arrived. My office and department were in the basement. At one point, I needed two buckets to catch the water leaking into my office during rainstorms. The HVAC system was outdated. In late Spring, they declared Winter was over and turned off the heat. We often got a snowstorm or two after the heat was turned off. I remember seeing the breath of students in coats and mittens during one of my lectures. Maintenance in this old building was also subpar. Under one of the stairs to the basement level, there was a growing dust ball more than a foot tall, which I sometimes called "Cousin It" (Adams Family). As my joke started to spread, "Cousin It" finally disappeared.

As I have gotten older, I realized there is a price to pay for being a creative type. I struggle with being a good listener. My mind tends to wander. Inspiration strikes anytime, anywhere. I am good at remembering people but not names. I try to work on improvement, but it is hard. When I first became a department chair, my dean sent me to "chair training school", a retreat in the Florida Keys. We took the Briggs-Meyer personality profile test. I found that I am an extreme extrovert. I can't help myself. I am always starting conversations with strangers. Texans tend to be this way. Sometimes those conversations lead to amazing connections. People with my personality type (ENTJ) is the Briggs-Meyer label) include Napolean and General Patten. It was refreshing to know that I had a strong leadership profile with a personality type that people wanted to follow. But I am non-violent. So, I found the military connection a bit unsettling. I guess I took that side of me out on the chess board.

Being an extrovert can lead to a mixed bag of experiences. While sitting in a famous dive bar in Manhattan, a woman sitting next to me started a conversation. She was clearly intent on taking me home with her. She was a bit younger than me, extremely aggressive, but not attractive. Her last name was Gotti and, yes, she was related to the famous Mafia crime boss. I got out of there as fast as I could and, thankfully, she did not follow me. I wondered what would happen if a guy did not perform to her expectations. A name popped into my head, Lorena Bobbitt. Google it if you don't know her story.

Back to the chairs meeting in the Keys, I met another new chair from a top 20 medical school. The Briggs-Meyer test revealed that I was the most extroverted and she was the most introverted of the 20+ attendees. I learned that because the group leader had us all sit in a U-shaped arrangement going from most introverted (her) to most extroverted (me). The speaker also encouraged us to get to know each other. I later noticed her sitting alone on the beach and asked if I could join her. I knew several faculty members in her department, which was a good intro. I was particularly curious about how someone who did not like being around people would ever become a department chair. As expected, she hated going to parties and was usually the first to leave. She hated dancing. She was my polar opposite. I later learned from a friend in her department that most faculty could not wait for her to leave. He noted that her strongest trait was simply avoiding everyone as much as possible. She just wanted to be left alone in her research lab. Why on earth would someone with her demeanor ever take a leadership position. Sometimes poor leadership is more about the sins of omission than the sins of commission. But I would argue that someone who is actively destructive is worse than someone who is largely absent.

Realizing that we are all flawed human beings, I try to pay attention to my shortcomings and remind myself not to let them interfere with normal day-to-day functionality. It is very helpful when writing a paper that I can become so immersed in writing that I often lose track of time. Sometimes I look up at the clock and see that 4 hours or more have passed while it feels like only minutes. I wrote the first draft of this book in 3 weeks (but final tweaking is taking much longer). I tend to be a dreamer who can easily envision where I will be in 10 or even 20 years. This helps me stay focused on achieving those goals. Lessons from the book "Good to Great" were helpful in that regard. Don't be detracted by the shiny object, stay focused on the long-term goal. I love nature and notice the smallest details that others may miss (like the bee-flower photo). There was a book title that caught my attention years ago. Unfortunately, I never got around to reading it. But the title was intriguing: "The A students work for the C students and the B students work for the government." I have noticed many examples of that in my life and learned to surround myself with the A students, who are incredible at providing the necessary details needed for my work. In turn, they are grateful that I am taking them in a meaningful direction with my ideas, translating basic science bench work toward improving patient outcomes for instance (translational research). It is important to try and make your personality type work for you while realizing that self-improvement should always be in the back of your mind. I am a work in progress. Aren't we all!

Shortly after I stepped down as anatomy chair and moved to Sioux Falls to start the cardiovascular research center, the USD president decided to build a new basic science building in Vermillion. It was intuitively obvious to everyone that this was the time to consolidate the basic and clinical sciences in Sioux Falls. Students would have been 100% behind such a decision. But students, the sole reason for schools to exist in the first place, were never consulted. The university president had decided to put the new building in Vermillion and it was not open for discussion. He was so entrenched in the idea that losing the med school would irreparably damage the main campus that logic abandoned him. The pediatrics chair and I were the most vocal in opposition to this stupid decision. Dean Talley quickly closed the discussion. I would later find out that he almost lost his job when he was chair of medicine for promoting moving the basic sciences to Sioux Falls.

While Chairman of Anatomy, I often suggested to Dean Talley that we partner with the hospitals in Sioux Falls to build a cardiovascular research center. That opportunity came up during my 5th year on the basic science campus in Vermillion. While it was a top 100 US hospital, the Sioux Valley Hospital CEO realized that it would be good to develop research if they were to continue climbing the ladder. Note, Sioux Valley was later renamed Sanford Hospital due to a billionaire's donations. The hospital CEO approached Dean Talley and asked if he had any faculty qualified to run a cardiovascular research center. We submitted a business plan two weeks later. I was counseled to keep the proposal modest, and it was ridiculously so. But I knew success would attract more funding and I could work with the modest budget. In the partnership, Dean Talley committed three faculty positions from my department, my position and two others. I later asked Bob: "How on earth did you get permission from the president to move faculty positions from the main campus to Sioux Falls?" He said: "I didn't ask. I just did it because it was within my authority as Dean and was an excellent opportunity. Then I apologized later." I would later come to realize that the president would never forgive the Dean, or me for this action. Thankfully, Bob knew how important expanding research to the clinical campus would be for the med school. He also had my back.

I had a wonderful time during my 5 years as chair in Vermillion. I inherited some amazing teachers. In fact, a member of my department won the Basic Science Teacher of the Year award every year during that 5 year-span. I arrived in SD with a new 5-year NIH grant. There was no other funding in the department at that time. By the time I left five years later, our department had 6 NIH funded faculty (out of 10). We had more NIH grants and publications than the other three basic science departments combined. The other chairs were not well-liked or respected by many of their department faculty and morale was low. It was another good lesson for me of "what not to do." They pointed out to me that NIH grants were a double-edged sword, prestigious but often a ticket for faculty to leave. That was certainly the case in their departments. But I never lost one faculty who obtained NIH funding. What a novel concept: treat people well, with respect, provide a pleasant work environment, support their growth and development, and they will be happy where they are. When they decide to start looking, you have likely already lost them.

I quickly developed my own leadership style. I believed that a chair needed to earn the respect of his/her faculty by setting a good example while providing strong leadership and mentoring. In my mind, I worked *for* them and not vice versa. This was an alien concept to the other chairs and a primary reason why they could not retain their best faculty. I believed that the main role of chair was to recruit high quality faculty and do what it takes to help them succeed. This included protection of research time, providing adequate resources, and advising them on grants when I could. While realizing that I may have limited knowledge in their research area, grant strategies are rather generic. Most importantly, I believe that faculty should never feel threatened or intimidated. Life and research are hard enough as it is. It is important to encourage collaboration and establish a pleasant and collegial working environment. There will almost always be people who don't get along for one reason or another. That is one of the most difficult things to deal with as a chair. But they often realize that disrupting a collegial atmosphere does not look good on their part and generally do their best to avoid open conflict.

Annual evaluations can often be stressful if one has had a bad year. Rather than come down hard on someone who has not had a productive year, I choose a different approach. I describe what progress toward promotion looks like. I point out what was done well. For example, a faculty may have had a good year teaching, did a good job on several committees and other service activities, but had an off year in research. I describe what successful research progress toward promotion looks like, the number and quality of papers, grants, and grant submissions (e.g. provide specific targets). I also realized you can't just look at one year in isolation because research projects can take years to complete. A faculty may have several papers ready for submission or under review at the time of evaluation. This should be a good start to the next year with a couple of early publications. I point out that if you average at least x, y, and z over 5 years, you should have a strong case for promotion. I give them specific minimal targets. I have often been surprised that new faculty may not have a good idea of specifics regarding attaining promotion. It is the job of the chair to help optimize their approach to career development.

I was fortunate to have inherited an excellent neuroscientist from Harvard. Luckily, his wife grew up in nearby Iowa, a short distance from Vermillion. So, this made retention easier. He grew up on the New Jersey shore across from Manhattan and could see the Twin Towers from his back yard. He bought a farm and enjoyed driving his tractor while learning how to "hobby" farm from his wife in his off time. He did not need much research mentoring and landed an NIH grant shortly after my arrival. Another faculty member was a single parent raising a mentally challenged child. Nonetheless, he was a devoted scientist that I would often see working late on weekends. He was almost ready to give up on NIH funding. He had not published a paper in 3 years. I found that he had a habit of abandoning good projects that were about 90% completed. I had him focus on completing these projects and pointed out that one only needs to show 2 or 3 good successive years of publications to convince NIH grant review committees that you are productive and your career is on track. I kept an eye on him, provided frequent encouragement, and discussed recent data that he was excited about. He met the goals I set for him, landed his first NIH grant shortly after making the adjustments, and was promoted to Professor. Another faculty I inherited had a similar situation. I worked with him and he landed his first NIH grant. I recruited two other excellent neuroscientists who landed large NIH grants after their arrival. We had 4 neuroscientists, 2 cardiovascular researchers, a prostate researcher, a developmental biologist, and 2 teaching intensive faculty with no ongoing research. Three of our four neuroscientists obtained NIH funding while the other was busy teaching neuroscience, gross anatomy, and embryology year-round with minimal research. He would later become Dean of Students. Although this position largely took him out of the classroom, he was a perfect match. I believe he remembered every student he ever taught and much about their families. I was amazed at how many past students kept in touch with him. It was a no-brainer suggestion to make to the dean. Find the best match for the job even though it may annoy some who coveted it.

I inherited several outstanding, award-winning teachers, who each carried the teaching load of 2 or 3 people. They were not competitive scientists but were clearly valuable to the medical school. After all, teaching medical students is really the most important activity at a medical school. I fought for greater respect of teaching excellence. One thing I stressed was that research-intensive faculty had protected research time because these outstanding teachers freed up time for them by teaching more. We were all part of a departmental team providing overall excellence in teaching and research. I made sure there was an atmosphere of mutual respect for everyone who worked hard. Most of all, academics who work hard just want to be respected and appreciated. For some, this may actually be more important than a pay rise. They also enjoy being part of a successful team.

I never lost one NIH funded faculty member while chair in Vermillion. When I moved to Sioux Falls and stepped down as chair to start the research center, that left 3 of the 4 basic science departments without a chair, with only the biochemistry department having a permanent chair. The Dean did not have enough money to recruit 3 chairs. So, he consolidated all 4 departments under the remaining chair. I warned that many in other departments do not want to work for this guy or his right-hand assistant. The consolidation chair had another faculty with whom he delegated much of the department work. She earned the nickname "Little Hitler", which says it all. Within a short time, half the NIH funded faculty departed. Quality leadership may not be everything, but without it, you will have only mediocrity. In most places, faculty who earn their keep by providing a lot of quality teaching but are not marketable because they lack a big transportable grant, often bide their time until retirement. Then they move away or stay if they like the place. I don't view these faculty as failures to be looked down upon. They help keep the place running by doing the primary job of a medical school, training physicians. Sadly, I don't know of any med school that is famous for teaching. The reputation of medical schools is largely based on their research.

For those who really enjoy working with amazing young people and training them to become physicians, this is a rewarding life even if background politics can be annoying. Is there any job devoid of politics?

I have many students who have stayed in touch with me for decades and it is so much fun to hear of their career accomplishments. Some of these students are so unique that you just can't forget them. The med student who got the isolated human myocyte prep working in my lab went on to become a leading HIV-AIDS researcher at the Center for Disease Control in Atlanta. He later became the AIDs Commissioner in New York City overseeing all related programs during the AIDS crisis when we still did not know the cause. He designed the well-known free condoms with NYC Subway wrappers. When I last spoke to him, he was spending much of his time combating the AIDs crisis in Africa.

I have pleasant memories of Vermillion. Only slightly larger than my hometown in Texas, it is a nice, small college town. It has a domed stadium, which hosted football, basketball, swimming, diving, and track and field at the time I was there. The Dakota Dome is also a very nice place to work out in the cold winter months. Just about everyone would be surprised to know that Vermillion also has a world-class museum, The National Music Museum. This museum has one of the best collections of musical instruments in the world. This ranges from "The Harrison", considered by many to be the best surviving violin made my Stradivarius, to the saxophone given to Bill Clinton by LA Sax. Since Bill could not accept the expensive saxophone as a presidential gift, Tom Daschle convinced him to give it to the museum. The museum has one of only four surviving guitars made by Stradivarius, the world's oldest playable harpsicord and cello, one of Elvis Pressley's guitars, and a wide variety of brass, woodwinds, and string instruments from around the world. The collection of instruments from Africa shows the amazing creativity and inventiveness of people from that continent. The museum, on the USD campus, offers the only graduate degree in musical instruments in North America. I would encourage anyone on a trip across the US to stop by this museum, especially if you are a musician. I would often see scholars from around the world coming there to study a given set of historic instruments. Visiting experts would sometimes give concerts with these instruments, often playing music of the period.

Success at the **CRC** (Cardiovascular Research Center) led the hospital CEO to raise the research investment to $10M shortly after its inception.

Little of the increased funding came to our center, likely because we were generating significant grant funding and had become largely self-supporting. This research success attracted the attention of the State's only billionaire, Denny Sanford. Sanford made much of his fortune in the credit card industry. South Dakota and Delaware had passed banking laws that are quite favorable to that industry as you may notice when you pay your credit card bills. Bill Janklow was governor at the time these banking laws were passed.

It is interesting how you can get to know influential government leaders in a small state. I spent the evening chatting with Governor Janklow at the St. Mary's father-daughter dance one night in Sioux Falls. Bill was there with his granddaughter, who went to school with my daughters. It says something about the importance of family for him to leave the Governor's office and drive half-way across the State to be with his granddaughter for this dance. His daughter was a divorced, single parent. You may remember him due to an incident that attracted national attention when he was the SD House representative in DC. He had diabetes and liked to drive fast on isolated South Dakota highways. He ended up running a stop sign going over 70 mph and killing a motorcyclist in his Cadillac, apparently related to his diabetic condition. He was convicted of manslaughter, sentenced to 100 days in jail, and resigned from Congress. It was the end of his political career at age 64. Bill was a force of nature. Our politics were different, but I always enjoyed my conversations with "Wild Bill Janklow." He had a human side as noted above, but I had no success in stimulating an interest in medical research.

Regarding the credit card business, Denny Sanford's card was a bit controversial because of the high fees and interest rates. Although First Premier Bank targets those with bad credit and many just dig a deeper hole, it also provides an opportunity to fix your credit when no one else would take the risk. Sanford donated $400M to our partner hospital, mostly targeting children's research and care. At the time, this was the largest single research donation to a hospital in the US. Other donations by Sanford would follow, including a $70M donation to the particle physics lab at the former Homestake Mine on the other side of the state. Regardless of what you may think of Sanford, his philanthropy changed the state for the better and certainly provided a catalyst for research. Creating a research success story in Sioux Falls played a key role in attracting his attention.

After moving to Sioux Falls, I often suggested to Dean Talley that we should start a MD/PhD program. Such programs are often critical in providing future leadership in medicine. It was clear that USD basic science leadership opposed this, believing it would provide more benefits to the growing research enterprise in Sioux Falls than to the basic science campus in Vermillion. Nonetheless, I had an opening to get this done. With growing research in the state, Governor Mike Rounds, who is now a US Senator, had some interesting comments in a special session in Pierre, the state capital. He lamented that South Dakota was dead last in PhD programs in the US. While in the restroom at the break, I said rather loudly: "Well that's a no brainer. There is one PhD program that would instantly attract national quality students." I said this loudly because there were members of the South Dakota Board of Regents at the urinals on both sides of me. They immediately asked what that program would be. I said: "A MD/PhD program, because the students had to be high enough quality to be accepted to med school too."

The MD/PhD program was announced a few months later. I believe initial State funding was about $500,000/yr. But leadership in Vermillion put someone in charge with the clear intention of sabotaging the program, "Little Hitler". She and her husband (wonderful guy who later divorced her and moved to Sanford Research) were recruited from an Ivy League school on the East coast. I heard that she agreed to come only if she did *not* have to teach medical students, only graduate students. Now, that's an interesting stipulation for a professor at a medical school! Failure to recruit a MD/PhD program director drug on for quite a while until USD was finally pressured by the Governor to "use it or lose it" regarding the appropriated funds. Eventually, the president made an unusual move clearly intent on punishing me and Dean Talley for moving faculty positions to Sioux Falls. He had heard that my top faculty member at the CRC was looking elsewhere because of the racist behavior at Sanford Research by the director of the research center who did not like Chinese. How could you do research in the US for the past 30 years without the influx of amazing talent from China? I complained about this behavior to no avail. The faculty member being recruited by the USD president was XJ Wang.

XJ was one of my previous PhD students at USD and got his medical degree in Wuhan, China. After a postdoc with Jeff Robbins, I managed to lure XJ back to help me build the research center, before the rest of the world found out how good he was. Jeff was very helpful with some arm-twisting. XJ quickly made a name for himself and landed two large NIH grants within a few years after returning to SD. The behavior of the Sanford Research director was making him feel unwelcome. By then, many schools were interested in recruiting him. He was frequently away on job interviews. But the USD president made him an offer that no one could match, he offered to double his salary and promote him from Assistant Professor to Professor with tenure, while naming him the director of the MD/PhD program, retaining the state funding at the last minute. We were both USD faculty and continued to collaborate. At XJs PhD graduation, I introduced him to the president and predicted that he would become the most famous PhD graduate from our school. XJ has now surpassed my research accomplishments regarding publications and grants. I am like a proud father. He has fulfilled my prediction. With XJs move, I became the only USD faculty employee at Sanford Research Center. The third USD faculty member I originally brought with me left to became Director of Research at the American Diabetes Association. That turned out to be a good connection for funding opportunities in diabetes.

In the early days of the research center, I mentioned to thc USD president that NIH funding was biased toward larger institutions. Smaller institutions were not eligible for the big $10M grants (**Program Project Grants, PPGs**) because they simply did not have the resources to put together critical masses for those grants. We decided to write our congressmen. Then Senate Minority Leader, Tom Daschle, was most receptive. This was shortly before Tom became Senate Majority Leader. We also made the case that this lack of NIH funding investment was a major roadblock to developing biotechnology companies in these states. The NIH Deputy Director responded that the NIH already has the IDeA program for underfunded states. Realizing that this was a pittance of funding, Tom wrote Donna Shalala, Director of Health and Human Services, which oversees the NIH (**Figure 12**. Copy of letter from Daschle's secretary, unfortunately cc'd before he signed it). The next budget (1998) contained a very large increase in funding for 23 underfunded, mostly rural states, and Puerto Rico. It was still less than 1% of the NIH budget but it made a big difference. There were some nasty complaints published in leading journals from scientists at top schools stating this was a huge waste of money. I met one of those guys at a scientific meeting later and asked a simple question: "Do you realize these 23 states have 46 senators who mostly voted against NIH funding in the past? How do you think this will influence NIH funding at your place?" The same budget increasing the IDeA funding also marked the beginning of doubling of the overall NIH budget over the next 5 years. Tom was hesitant to take credit. But there is no way in hell that NIH would have done this without pressure from Congress, triggered by one of its most influential senators. The people in charge at NIH were largely under the influence of those who opposed this as noted by their nasty letters to major journals. **The lesson here is that a well-placed letter to the right person can sometimes make a huge difference. Nothing ventured, nothing gained.**

The first year of the NIH IDeA expansion contained a PPG like grant program as proposed to Tom Daschle in my letter to him. These $10M **COBRE** (Center of Biomedical Research Excellence) grants were initially limited to only two per state and USD politics prevented my center from applying in the first round. The politics were against us because the USD president was not happy that the CRC even existed. Ironic that my letter to Tom triggered establishment of this program but my group was excluded from submitting. The neuroscience group assembled during my tenure as anatomy chair landed one of these grants on first submission. The other proposal submitted by South Dakota State University was rejected. The IDeA expansion also contained a program called IDeA **INBRE** (Networks of Biomedical Research Excellence) aimed at building research capacity primary at undergraduate institutions. It would have been difficult for my group to submit a COBRE proposal in the first round anyway. I came down with a head and neck cancer a few months before the due date, and we were short a project. Proposals needed 3-5 projects and an NIH funded Principal Investigator. Luckily, I survived major surgery and radiation treatment. After making the case that SDSU had their shot and it was our turn, we managed to get the green light the next year. Our $10M COBRE grant was funded! With this infusion of NIH funding and several more major NIH individual grants, our CRC was becoming very successful. There were no data to compare CRCs across the US. The closest comparison would be physiology departments. If we were a physiology department at that time, we would have ranked #2 in the US in funding per faculty member (~$700,000 per faculty per year), behind Johns Hopkins. Of note: We later managed to get the COBRE grant renewed for a total of $20M over 10 years.

THOMAS DASCHLE
SOUTH DAKOTA

COMMITTEE

AGRICULTURE

(202) 224-2321
TOLL FREE 1-800-424-9094

Email: Tom_Daschle@daschle.senate.gov
Internet: http://www.senate.gov/~daschle

United States Senate

WASHINGTON, DC 20510-4103

20 6TH AVENUE, SW
SUITE B
P.O. BOX 1536
ABERDEEN, SD 57402-1536
(605) 225-8823

816 6TH STREET
P.O. BOX 8168
RAPID CITY, SD 57709
(605) 348-7551

320 NORTH MAIN AVENUE, SUITE B
P.O. BOX 1274
SIOUX FALLS, SD 57101-1274
(605) 334-9596
TDD (605) 334-4632

September 22, 1997

The Honorable Donna Shalala
Secretary
Dept. of Health and Human Services
200 Independence Avenue, S.W.
Washington, DC 20201

Dear Madam Secretary:

I was contacted recently by Dr. Martin Gerdes of the University of South Dakota (USD) regarding his concern that the National Institute of Health (NIH) does not distribute research dollars equitably. I have enclosed a copy of that correspondence for your review.

As you will note, Dr. Gerdes asserts that smaller schools and research facilities are not enabled to compete for NIH research grants. Dr. Gerdes notes that NIH often targets its grant funding to larger institutions, while smaller research facilities become less and less able to compete. Dr. Gerdes recommends offering grants to help smaller schools who are seeking to become more competitive in a specific specialty.

In addition, I also received a letter from James Abbott, the President of USD, who enclosed a research funding levels list that shows the enormous amounts of grant money schools such as Harvard and Johns Hopkins receive in comparison to USD and other smaller schools. Dr. Gerdes suggests that smaller institutions may actually be more efficient in the way they spend their federal research dollars.

I would greatly appreciate your response to the points made in Dr. Gerdes' letter and reinforced by the correspondence from James Abbott, as well as any additional information pertinent to this matter. I look forward to hearing from you.

With best wishes, I am

Sincerely,

Tom Daschle
United States Senate

TAD/stn
Enclosure

PRINTED ON RECYCLED PAPER

Figure 12: Letter from Senator Tom Daschle to HHS Secretary Donna Shalala

Our success attracted the attention of Governor Mike Rounds. He established **South Dakota 2010**, which included research grants patterned after our CRC critical mass. We were eligible because I was a state employee and was qualified as principal investigator and a project leader because I obtained no direct funding from our NIH COBRE. COBREs were directed to funding new investigators but needed a senior level NIH funded leader (me in this case). We landed the first of these 2010 grants for $4.5M over 5 years. So, when I came to SD, research was a 4-letter word. "We were paid to teach," I often heard. And now, we landed a multimillion dollar grant from the state's Republican governor. Wow! What a sea change! For the 2010 proposal, we could not fund our already NIH COBRE-funded CRC group members. I solicited projects from several Sanford Research investigators who were not on the COBRE grant. The pickings were slim at that time, the Sanford Research Director (yes, the racist), a cancer researcher with limited publishing success, and a new recruit to the CRC who was brilliant but struggled to publish. The Sanford Research Director had just completed a 5-year NIH grant with literally no productivity. For the next 5 years, I had to apologize for his continued lack of productivity in my annual 2010 Grant reports to the Governor's office at the state capital in Pierre. As I departed for the meeting, he would say: "Now don't go bragging about yourself." I accounted for 80% of the grant productivity and had to highlight our accomplishments. So, what was I supposed to talk about? I spent much of the 5-year period apologizing for the 3 other project leaders who always had "exciting" work ongoing but rarely published.

Near the end of the 2010 grant, I confronted the Sanford Research Director regarding what was clearly racist behavior regarding our Chinese investigators. His response: "Well if they think I am a racist, they better have damn good evidence, or I will sue the hell out of them." I immediately thought to myself: Wow, the appropriate response would have been something like, "Oh no, I am not racist and need to speak with them, find out why they feel this way, and correct it." He continued to harass other Chinese, including threatening to fire my very productive post-doc from Beijing who inadvertently missed a rat hair when cleaning up a surgical area. He also blamed a pregnant Chinese post-doc from XJ Wang's lab for a radiation spill, when it was a white tech responsible for the accident. What had been a very pleasant environment had turned rather hostile and was going to get worse. It is specifically spelled out in NIH documents that Chinese scholars are a protected group. Imagine how they would be treated if they were kicked out of a US lab for confronting their supervisor? How would they be treated upon return to China for defying leadership? They were afraid. I could have requested a NIH investigation but that would have been very painful, and the targeted Chinese begged me not to do it. Additionally, NIH grants might be frozen at the research center, affecting everyone including other Chinese investigators. **The lesson here is to avoid recruiting a narcissist. But if you do, make sure they are never put in a position of power and authority.** Unfortunately, these people are often very good at hiding who they are until it is too late.

When we had the grand opening of our CRC, it was a big deal for South Dakota and marked a partnership between the state's medical school and the leading hospital (Sioux Valley Hospital at that time). I wanted to get a big-name speaker. So, I managed to get a face-to-face meeting with the Director of the National Heart Lung and Blood Institute at NIH, Claude Lenfant. One of the first questions he asked me was: "Oh, South Dakota, does North Dakota have a medical school too, or are you guys combined?" At the time there were 123 medical schools (MD schools) and yes, we had separate med schools. It was clear from his comment that we were merely flyover country in his mind. This also confirmed my belief that the NIH had little interest in the smaller med schools located mainly in rural states. I was disappointed at the naivety of his question and was pretty sure he would say he was too busy to come to our grand opening. So, I told him he could pick the date, and we would plan around his schedule, giving him a 4-month window. I could tell he was a bit irritated at my outmaneuvering him, but he said yes. I also invited a world-famous heart failure researcher and cardiologist from Harvard, Beverly Lorell. Everything went very well. After visiting our partner hospital, Beverly was particularly impressed with the quality of cardiac care at this remote location. "Wow, they have everything we have in Boston," she said. I should say that the quality of health care in Sioux Falls is one reason this city often makes the list of best places to live or retire.

I believe that Claude also enjoyed the visit. During a conversation with him, I made a request. First, some background information would be helpful. The lower the score, the better the NIH grant. There is automatic funding of grants below the "pay line", which is about 15% these days. But the success rate is a few percentage points higher because NIH officials select other grants that just missed the pay line. Many things come into consideration here. For instance, if they already fund many grants in a specific area but an important area is underfunded, they may select such a grant on the bubble. Promising minority investigators also fall into this area. At the time, I had a grant submission that soon received a score within that bubble but just missed the pay line. Claude told me to let him know if this happened and he "would see what he could do." Our grant was also in an important, underfunded area and was critical for our fledgling new research center. He selected the grant for funding. Of note, this is a grant that funded a paper that world leading heart failure cardiologists have told me is among the most important papers ever published in heart failure. I say that to highlight that there was excellent science in that grant proposal. Funding limitations make you wonder how much outstanding science goes unfunded and the important work that never gets done. Claude knew we were serious about building a competitive center and recognized that it was unusual to see such local investment in research out in the middle of nowhere. There are lessons here. Be alert to opportunities and pounce on them before they are gone. Principal Investigators should also try to know the administrator overseeing their grant submissions at the NIH. They often give very helpful advice on revisions. Most importantly, he/she will often be involved in selecting bubble grants for funding. Psychologically, the administrator may be more favorable to someone he/she knows and likes when selecting from a group of grants with similar good scores but from unknown investigators. There are lessons here. When going into an important meeting, be prepared, anticipate questions and have a winning plan. I knew Claude would never have visited SD had I not met with him in person and anticipated key questions he might ask. Also, understand the system you are working in and how it can best help you.

The CRC grew to ~50 people, including faculty, post-docs, students, technicians, and support staff. On several occasions, someone would walk into my office and tell me this was the best place they have ever worked. It was an exciting, productive, and happy place. Heck, I even had my own personal restroom and an office with a view. Eventually, I became the only USD employee at Sanford Research while we were clearly moving toward a "divorce" between the hospital and the university. The hospital CEO had put forward an "alimony" payment proposal to complete the divorce. This was $500,000/yr ongoing. The medical school desperately needed this money to fund research startup packages for new faculty recruits. The president had used another Sanford $10M gift for med school renaming rights to replace the roof on the Dakota Dome. So, who knows where that money would go. I sent the president a few emails warning that the offer would soon be off the table based on what I was hearing at the research center. He never responded. Eventually, the hospital took sole control of Sanford Research and USD got nothing. I shared my frustration with one of the president's administrators on a plane returning to Sioux Falls and it got back to the president. Big mistake. The thin-skinned president had just run off an amazingly competent chemistry professor who became one of USDs best administrators and made a mark as president of EPSCOR, a national research organization. The chemist moved to the University of Montana where he later became president. I heard he was unanimously drafted for the job by the faculty; they were so impressed with him. The USD president did not like having someone around who was far more competent, well liked, and likely viewed as a better replacement. As he departed for Montana, he sent me an email saying: "I don't care what the president says, everyone knows that you always have the best interests of USD in mind."

A short time later, Bob Talley retired as dean and was replaced by someone more compliant. Rod was a nice guy but clearly met the president's selection criteria. He once showed me a marionet of himself behind his office door and indicated the president controlled the strings. He hinted that the President would be coming after me even though I had tenure. Bob was a very strong dean and had no marionet. The president was afraid of him and gave him clear reign for the most part. Shortly after Bob's retirement, the divorce happened and I no longer had the backing of a strong dean to deflect the ill intentions of the USD president. I was pressured to leave USD and join the hospital, whose leadership did not like having someone in a position of power within their organization who was not under their control. Importantly, USD had a nice pension that I would be stupid to walk away from just before being vested. I continued to confront the racism by the Sanford Research Director and fought for improved benefits for the research employees. With the split, the USD benefit package was replaced with one much worse as Sanford Research employees. The divorce was also painful for them too. Eventually, Sanford sent a "goon" to my office who warned me to keep my mouth shut. It was clear that academic freedom was an alien concept to our partner hospital. It was the beginning of the end for the wonderful cardiovascular research center I built. Many workers came to my office to express their shock.

Several years after my departure from Sanford Research, they were unable to fill my director position and most of the Chinese employees at the CRC had left. They eventually closed the CRC, the catalyst for all the basic research development at Sanford Research and South Dakota. I heard that their only strong applicant for the CRC director was Chinese. That was a no go considering the Sanford Research Director's opinion of foreigners who were not white. Muslims were not popular with him either. He often joked the proper solution to the Middle East problem was to "nuke the whole area, turn it into glass". I asked him if he was joking. He said he was serious, and he was. Another lesson here. Pay attention to what people say in jest. It often reveals who they really are. In addition to being a racist, he was a narcissist who was extremely envious of my success and that of XJ. We have learned in recent years that narcissists are the worst leaders. They run off all the best people. Experts advise that the only thing to do with a narcissist it to turn and run away as fast as you can. Sanford Research eventually grew tired of his bullshit and fired him several years later. Better late than never. Fortunately, the research center I started has grown, thanks to the billionaire's endowment.

Soon after being relieved of my appointment as CRC Director, I moved back to our original research space at the med school where I had a tenured Professor position. Without the administrative work, it was a very productive year for me. I landed another large NIH grant, an American Heart Association grant, and a grant from the American

Diabetes Association. As I advise my faculty, make sure you stay productive and marketable. It not only means you are a keeper, but you have an off ramp if things go south. Regular faculty positions are generally stable and less affected by politics. If you work hard, make tenure, and remain productive, a lifelong job is the usual outcome. I thought I had the most secure position imaginable as the CRC Director who launched a major research success story that spread throughout the State. But I was caught between hospital and university leadership who simply could not get along. Leadership positions at med schools, such as department chairs, pay well but can be precarious since you serve at the pleasure of the Dean and they come and go with a high frequency. The tenuous nature of these leadership positions is why chairs also are generally co-appointed as tenured professors and can revert to these positions when they step down from a leadership role. Nothing in life is certain. I wish Sanford Research well. For some at Sanford, particularly those who struggle with NIH funding, the billionaire's pool of money is too hard to walk away from, golden handcuffs. Just don't rock the boat. Readers may be surprised to know that politics can play such an important role at med schools. Hopefully, there may be a few helpful lessons here for those already involved in medical school administration or mid-level leadership in general. Keep a close eye on those who envy your success. And if they become your boss, it may be time to look for other options.

I had created a major research success story twice now, as anatomy chair, and as the founding scientist who started Sanford Research and set the stage for nearly half a billion in private funding and well over $50M in NIH funding. Not bad, when considering South Dakota struggled to obtain even $1M in annual NIH funding before I arrived. I was there for 19 years and we moved past 19 other medical schools in NIH funding during that time. It is a wonderful legacy that still puts a smile on my face. I never expected to end up on Mount Rushmore. I knew I had to rattle the cage to make a difference, and that could be dangerous. But I would not change a thing.

I have a couple more stories related to politics. In the first story, I had an opportunity to save the planet but failed. You win some, you lose some. I often took my family to Saturday brunch at Sanaa's in Sioux Falls. Sanaa, who grew up in Syria, was a master of middle eastern cuisine. She was married to former US Senator Jim Abourezk, who I became good friends with. Jim was the first Arab-American to serve in the US Senate. I loved listening to Jim's stories. One story involved negotiating an agriculture agreement with Russia. In his earlier life, Jim was a bartender and bar owner. His ability to hold his liquor came in handy as he negotiated a farm agreement in Moscow. Before the Russian representative would even discuss agriculture, they had to finish a bottle of vodka. Jim held his own and the negotiations were a success. His Senate colleague along for the discussions was a teetotaler and was completely ignored.

Now the story about failing to save the planet. A few weeks before the 2020 election, Jim waved me over to his table as I arrived at Sanaa's with my family for Saturday brunch. Sitting at the table was Tom Daschle, who was about to become the new Senate Majority Leader. I thanked him again for the role he played in NIH funding increases. Tom, surprisingly humble for a politician, gave credit to others who got it done. Also sitting at the table was Ralph Nader who was running for president as a third-party candidate. This was a good opportunity for me to just listen and learn. So, I did. As we left, I begged Ralph to consider dropping out since the polls showed he had no chance to win but would pull enough votes away from Al Gore to give the election to George W Bush. That's exactly what happened, of course. Ralph just looked at me and smiled. I was unable to penetrate his ego. As we all know, a major platform of Al Gore was saving the planet. He even wrote a book about it. Although I am sure Daschle also discussed the topic with Nader, I had an opportunity to save the planet but failed. I am also saddened by the black mark this left on Ralph Nader's legacy. He contributed so much to consumer safety and protection.

I also got to know the current (2026) Senate Majority Leader, John Thune. Like me, John is a small-town country boy who moved to the city. John was a consultant for Sanford, our partner hospital, before being elected to Congress. Shortly after the 2000 election, things were going so well for research development at Sanford Research that the CEO decided to throw a big party for his employees. Sanford was well on the way to becoming one of the largest rural health care systems in the US with many branch hospitals in several states. There were thousands of Sanford Hospital and Research employees at the party where Sheryl Crow and others provided entertainment. It was open only to Sanford Hospital employees and privately funded by the billionaire's money we were told. To the excluded locals, this came across as rather extravagant, leading to a negative article in the local newspaper. But for us it was a wonderful day when I spent much of my time in the VIP tent visiting with John Thune and others. He would later defeat Tom Daschle for the Senate in 2004. The TV ad that defeated Tom was misleading, politics as usual. The ad warned South Dakotans that they would lose the family farm due to proposed increases in the "Death Tax" (Estate Tax). I remember thinking that it would be more convincing if they had someone in the ad who actually lost their farm because of this tax. In fact, I later read that Bush asked the National Farm Bureau to find such a person. The best they could do was identify heirs who lost the farm due to bankruptcy but were unable to find a farm lost due to the Estate Tax. Oh well, the message worked and South Dakota voted then Senate Majority Leader, Daschle, out of office. Little did I know at the time that he was being replaced by a future Senate Majority Leader. During his early tenure as Senator, I often ran into John at the Minneapolis airport where we were waiting for the connection to Sioux Falls. With his likeable personality, I can see why he became so popular in his party and rose to Majority Leader. South Dakota is an interesting state where regular people can get to know their governors and congressional reps on a first name basis. I did my best to make it work for things that mattered.

Chapter Eleven
From South Dakota to New York

In 2010 after moving back to USD Medical School, I began to apply for other jobs. I had many interviews, but one was particularly interesting to me. My success in SD caught the attention of the DO school at New York Institute of Technology **(NYIT)** on Long Island (New York Tech). Doctors of Osteopathy had established themselves as equals to MDs in the medical community. But few DO schools across the US were contributing to new medical knowledge. NYIT was among those who wanted to change that. On the evening of my interview, I was dining with Dean Tom Scandalis at Rothman's Steakhouse, a famous restaurant that was the favorite of Elizabeth Taylor and Richard Burton when they filmed in the area. Tom got right to the point. "We can talk about baseball (both fans) or whatever it will take to get you here?" The med school at **NYITCOM (NYIT College of Osteopathic Medicine)** had very little research going on. I sometimes joked that: "I don't think anyone there at the time could even spell NIH." They once had a faculty who landed a big NIH grant, but he quickly left. There were three major obstacles for research development at NYITCOM: 1) few PhD researchers knew what a DO was at the time; 2) there was very little research infrastructure (e.g. established grants management, cutting edge equipment); and 3) how could I attract NIH competitive faculty with literally nothing to show them but promises. In short, it was the perfect job for me, and they knew it. When I met to negotiate the position with the Vice President of Health Sciences, Barbara Ross-Lee, she offered a higher salary than I suggested. They were serious. You may know Barbara's sister, a well-known singer. According to Barbara, Diana Ross may be a better singer, but she was a better dancer. It must have been cool singing in the church choir with young Diana Ross. But my interview was poorly timed. The week before, Diana Ross was playing next door at the Tilles Center. Who knows? I might have been able to tag along with Barbara to meet her sister.

On my interview at NYITCOM, I noticed some of the happiest med students I had ever seen. There was clearly a good rapport between students and faculty. I also noticed the location was spectacular, a place I had always dreamed of living. Located on the North Shore of Long Island, the Old Westbury campus is like living in a beautiful forest. It was formerly the DuPont estate and has a beautiful "Great Gatsby" style mansion.

We initially rented a place on Cove Neck Road in Oyster Bay, about a 5-minute walk to Sagamore Hill, Teddy Roosevelt's estate. The North Shore is surrounded by water, rolling hills, and thick forest. European explorers, who settled the area in the mid-1600s, must have felt like they found paradise. There are beautiful natural bays along the North Shore; Oyster Bay, Huntington Bay, Cold Spring Harbor, Port Jefferson Bay, etc. The North Shore forms the southern boundary of the Long Island Sound with New York and Connecticut across the water forming the northern boundary. I can only imagine how abundant wildlife was at the time of early settlements. I often saw herds of deer in my front yard and sometimes needed to brake for deer on my drive home. We later bought a house in Cold Spring Harbor, then moved to Huntington Bay. The commute was nice with water views along the way and rolling hills filled with dense forests. While housing in North Shore towns are typically sized lots, in between there are many estates with very nice mansions covering many acres. Even smaller homes often have 2 or 3 acres. It is a beautiful place to get lost on a Sunday afternoon drive.

When showing around potential recruits, I enjoyed driving them around this beautiful area. I sometimes posed a multiple-choice question when driving on the street where I rented a house upon my arrival to NY. There were three billionaires living at the end of Cove Neck Road. Of a cluster of seven mailboxes, I asked them which two belong to billionaires? They were all just plain mailboxes. Some recognized the name on two and answered correctly. You could see the home of the third while walking to the beach front at Teddy Roosevelts estate. It could have been the Chinese embassy as it sprawled up the hill. It belonged to Charlie Wang who passed recently. An added perk was seeing Billy Joel's incredible mansion across Oyster Bay on Center Island. There are more billionaires over there. Let's see, where do you live when you can afford to live anywhere? Oh, my neighborhood! Thankfully there are some more modest homes too. I loved kayaking and walking here.

Barbara Ross-Lee became dean again after the departure of Tom Scandalis. I got to know and respect her over the years. She was a military doc and believed in following the rules. But I found her to be open-minded if I successfully explained why a given rule should be changed. For instance, the NYITCOM rule for married employees was that they could not work in the same department. Often married scientists work in the same department or even the same lab when they meet. They often become life-long collaborators on grants and publications. It can be a plus to recruit a high-quality duo to a department because it can be difficult for both to land positions at the same place and time, thus improving retention. I had this situation twice during recruiting. I was able to convince Barbara to change the rule. The new rule was typical of elsewhere, you can be in the same department, but spouses cannot supervise one another. I once tried to explain my approach to building research at NYITCOM, which proved to be very successful. Knowing that research was not her strength, Barbara said: "I recruited you to build research. I trust you." I really enjoyed working with her. I simply explained what I wanted to do and why. She listened and did what she could to help. Since arriving here 15 years ago, I have been through 5 deans, which is surprisingly common. As would be expected at a DO school, none had a significant research background. A good leader who doesn't know about something important in their realm, hires qualified and competent people who do, and trusts them. Those who think they know everything and ignore good advice from knowledgeable people can do a lot of damage. Thankfully, I had enough support to build another research success.

My research approach is largely based on what young people want and need to succeed in science. First, I should explain where the US is at this time regarding recruiting young people into science. When I was in graduate school, almost everyone was US born. Since then, fewer Americans have been going into academia, including biomedical science research. Starting in the 1980s, probably 75% of typical PhD programs are now filled with Chinese students and those from other countries like India. Chinese students looking for opportunities at US med schools are typically very intelligent and sometimes have medical degrees and prior research experience. They typically make straight A's in their coursework. Their system, however, has stymied creativity to this point. It is interesting to note that Nobel Prizes have been awarded to Chinese *after* immigrating to the West, but rarely to those who never left China. These transplants have been critical for the US to retain dominance in research. These days many have become leading scientists here, often running departments. But more of these excellent trainees are returning to China these days. As a result, China is now catching the US at a very rapid pace. They will likely soon surpass the US in research productivity and patents. We have long needed to make research more attractive to our youth. My generation was loaded with workaholics and certainly Chinese immigrants have adapted to that model very easily. It is critical that we keep our door open to immigrant scientists.

Young people these days also want to have a better work-life balance. For the sake of mental health and happiness, I think they are right. This is largely the reason I gravitated away from the top schools. The intensity works for only a select few and they often burn out. At the upper third of med schools (ranked by NIH funding levels), there is extreme pressure to have at least half your salary on NIH grants. You may even lose your job if there is a continuing lapse period in grant funding. So, your success often depends on all 3 assigned grant reviewers supporting your grant. You can have 2 reviewers who love your grant and another who may be hostile to the specific research area, irrespective of the quality of your proposal. Sometimes the negative reviewer is biased and should not be reviewing that grant. This is the person who keeps an excellent grant from being funded. I would argue that the NIH should have a group of full-time paid grant reviewers at the core of study sections. These would likely be experienced scientists near the end of their career who have a lot of career experience with NIH grants. At present, the review system is virtually free. Some study section reviewers may spend 3 months a year reviewing grants but only get reimbursed for travel expenses for the most part. It can be a valuable experience for active researchers to see what is being funded, firsthand. But this is added to their already overloaded schedules and does not help much with promotion. There is even more pressure at top schools to have multiple NIH grants. Imagine how successful these researchers would be if they had more time to do actual research rather than spending 3-6 months of the year constantly writing, submitting, revising, and resubmitting their own grants while serving on study sections reviewing the work of others.

Excessive grant writing also contributes to reduced publication rates. Investigators need to produce a broad array of preliminary data for grants. The data collected is typically less than what is needed for a research paper. For an experimental group, data sample size of 3 is good for a grant but 6+ is typically needed for publication. So, instead of focusing on the data needed for publication, investigators often have a lot of unfinished projects. Excessive grant writing is stressful and may provide an impetus for fudging data to make a grant more attractive, particularly if your job and home mortgage are on the line. There should never be an incentive for cheating in science.

I believe the mindset in the 3rd quartile of NIH funded medical schools is a good place to be. Most have solid hard dollar salaries, particularly after promotion to Associate Professor with tenure. Those in the bottom quartile may just be small, located in less desirable places, or have uninspiring leadership. Of note, DO schools generally provide hard dollar salaries for all basic science faculty. But teaching loads tend to be higher at DO schools since they typically have less than half the number of basic science faculty per student as MD schools. The bar is higher for Professor, the next rank above Associate Professor, and often requires significant extramural funding. Large NIH grants are preferred over lower amounts from foundations like the American Heart Association. A national or international reputation in your field of research is also generally a criterion to make Professor rank.

Doing a large amount of high-quality teaching can often lead to employment retention, but retirement without reaching Professor is the usual outcome. I believe that teaching deserves more respect than it gets. After all, the primary mission of a medical school is training (teaching) physicians. Having one NIH grant most of the time should be acceptable and enable a respectable level of productivity, particularly if they have good collaborations. With less pressure on surviving and worrying about the stability of your position, stress levels should go down, and science should become more fun. It is also helpful to have some faculty who have not been very successful with grants pick up more teaching and service to free time for the star researchers to be as competitive as possible. This may be more important at DO schools. The higher number of faculty at MD schools to cover a given curriculum puts more pressure on obtaining grant funding to cover salaries. They simply do not have the funds to cover all basic science faculty salaries, especially at the more prestigious med schools. One of the downsides of the NIH budget doubling was that many schools overrecruited basic science faculty and are now struggling to make the payroll as the NIH funding pay line has declined. Of course, we really do need to increase NIH funding to ensure that all meritorious projects come to fruition. Despite what can be a rollercoaster ride, pursuing research you are passionate about is amazingly rewarding. And interacting with bright student researchers and sharing in their excitement of discovery is just plain fun.

When taking the Biomedical Science chair at a DO school, I was initially concerned that there would not be enough protected research time for research intensive faculty to obtain grants. On average, allopathic schools (MD) have two or three times the number of basic science faculty per student as DO schools. So, this was a legitimate concern since I was charged with building a strong research presence in the department. Luckily, that did not turn out to be the case. My department has 20 faculty and teaches over half of the 2-year basic science curriculum to 600 first- and second-year students each year. Nonetheless, 12 of my faculty have obtained an NIH grant and about 70% are NIH competitive. Having said that, the younger generations are different from my generation and I believe we need to cater more to their desire for work-life balance and make doing science more attractive as a career option. This is even more critical now that the immigration pipeline has come to a standstill recently.

It is also a bad idea to assign specific duties to someone that do not match their preferences or skillset. Chairs should do their best to match departmental needs with individual talents and passions. If you love what you do, you are more likely to do it well. Besides teaching and research, there are important service duties, like committees. Some of these committees can be rather time-consuming and require keeping up with changing regulations, but they are important. For example, the **IACUC** (Institutional Animal Care and Use Committee) must approve all animal research protocols, and the IRB (Institutional Review Board) must approve all human experiments. I believe a chair should look at all departmental duties and match faculty who best fit a given duty while ensuring a reasonable workload and protected research time for those who are grant competitive. For this reason, I often ask for volunteers to serve on committee since I can be surprised by who really wants to be on a given committee. A goal of the department should be to provide overall excellence in all areas of responsibility. This is something that does not differ from the business world. In fact, one of my favorite books is "Good to Great." This book describes what it takes for a business to go from good to great. When I interviewed for an endowed chair in cardiac surgery at the **Ohio State University (OSU)** School of Medicine, an associate dean was passing out this book to leadership and gave me a copy. Since then, I have shared the book with many of my faculty members and they have often said the book provided important advice relevant to running their labs and careers.

I made a rather humorous mistake during that visit to OSU. It was the Friday before game day and everyone was wearing scarlet and grey OSU attire. Many of the women working in offices even had OSU purses. I wore a blue blazer, yellow shirt, and blue tie with khaki pants. For those who don't know, blue and yellow are the colors of Michigan, the archrival of OSU. After receiving many glaring stares, I concluded that people in Columbus were not friendly, before realizing my mistake. They were playing Michigan that weekend.

Many studies also show that excessive time spent at work does not necessarily lead to greater productivity. Some recent studies even support moving from a 5-day work week to 4 days, allowing an extended weekend break with no loss of productivity. I have often noticed that researchers who leave after 7pm are often not correspondingly productive. I pay particular attention to those faculty and try to understand why they are inefficient. One of the problems I often notice is failure to tie up the loose ends of a project and publish. They often have projects near completion but move on to other projects that catch their attention. I suggest they post a list of all their ongoing research projects in plain sight and tell them to look at it frequently with the intention of identifying projects that could be completed and published in a short time. It is a wonderful feeling to cross off a study on the list because the associated paper was accepted for publication. There are books about the power of making lists. As far as daily chores, I have often noted that 4 of 5 can be done quickly. I do them first. Starting the day with a sense of accomplishment portends a good day.

When taking a leadership position at a new place, it is good to find another senior level leader you can trust and confide in. Ideally, this can be someone who not only knows where all the skeletons are buried but who can offer insight into the personalities and tendencies of people you must deal with. I was fortunate to find the perfect person when I moved to NY and we quickly became good friends. His insight was invaluable and saved me some grief.

It is also important to identify your own weaknesses. I tend to make decisions quickly and get things done. As a new leader, this trait had a bad result early on in one instance. I did not fully understand the hurtful repercussions to one individual of an important decision I made. I quickly learned to not only think things through but discuss the issue with affected parties beforehand. And, of course, it is always good to get buy in from those affected before moving forward. People don't like to be blind-sided.

I have been a workaholic my entire life. I love what I do. In fact, I don't think I had a real vacation until I was almost 50. After surviving head and neck cancer surgery and radiation, I finally thought it was time. Of course, I have had many invitations to speak around the world at beautiful places and often took a few extra days off. This is a nice perk since costs are typically paid for the meeting, including my airfare and hotel by the sponsor, my institution, or a grant. For the cost of a few extra airline tickets, the family can enjoy all the amenities of a wonderful place while the scientist attends the meeting. And a few out-of-pocket hotel nights turn it into something resembling a vacation. But I never had a real two-week planned vacation at that point. Of course, my wife hated flying, so I mostly traveled to these wonderful places by myself. It got lonely over time as most of my friends from the early days got married and brought their families, so we stopped hanging out at the end of the day. Of course, being an extrovert and alone is just another opportunity to meet new people.

I had cancer surgery in January of 2001, followed by many weeks of radiation. I needed speech therapy for several months to recover my ability to articulate clearly. I had 3 speaking invitations that July and asked my head and neck surgeon if I would be able to deliver those talks. He assured me that I would. I hit that goal about 2 weeks before the meeting. As I began my talk, I spoke of how grateful I was to be there. I was lucky to be alive. Everyone rose and gave me a standing ovation for several minutes. There were about 500 attending my talk in that session, one that typically drew about 100 people. This was my community and it brought tears to my eyes. I must admit that getting back to a job I loved gave me an additional purpose during recovery. There was much left for me to accomplish in my career. So much to be grateful for.

In my early career, drug companies showered leading cardiologists, particularly those organizing cardiology meetings, with many perks, including free hotel stays at leading resorts. This is now an illegal conflict of interest. But it was fun while it lasted because I was often on the list. My favorite meeting spot was the Boca Resort built by Addison Mizner in 1926 in Boca Raton, FL. I was among early leadership of the Heart Failure of Society of America when it formed, one of only a few PhDs. Luckily, my family came when I was housed in a wonderful suite overlooking the Atlantic at the Boca Resort. It had 2 very large, separate bedrooms with private bathrooms, and a large entertainment area in between. Most of my wife's family lived in FL and one sister was only a few miles away. So, it was a great opportunity for our daughters and their cousins to get to know each other. I guess this counted as a vacation, at least for the family while I was busy at the meeting. We spent a few extra days at Disney before the meeting and in the Florida Keys afterwards. I can still do this but the meeting costs have shifted to my grants.

The key to my success at the CRC in Sioux Falls was to recruit 2 of my best South Dakota PhD trainees back after they completed their postdoctoral fellowships. I spoke of XJ Wang earlier. I also recruited another former PhD student, Faqian Li back to my group as a faculty member. When these two superstar researchers were in my lab at the same time, it was a very productive period for me. They both attended one of China's best medical schools in Wuhan where they were classmates. At the time, XJ had already achieved Associate Professor very quickly and was considered among the top scientists at Hubei Medical School in Wuhan. Fortunately for me, early Chinese researchers felt they needed to get a US degree since the quality of Chinese biomedical research was not well established at that time. More recently, Chinese scientists with a doctoral degree equivalent realized they could start as postdocs. Their system is different than ours and this caused some early confusion. Their medical training is more like Europe where students come straight from high school and spend 6 years in training before getting a BS in Medicine, which was later realized to be equivalent to our MD. So, most US immigrants later put MD after their name even though the degree is a BS in medicine in China.

I was able to recruit both XJ and Faqian back to the CRC in Sioux Falls and they were key to our successful NIH COBRE application which resulted in $20M in funding over 10 years in addition to other funding. While working in my lab as a graduate student, Faqian had one of the most highly cited papers ever published in the Journal of Molecular and Cellular Cardiology, a leading cardiovascular journal. He also managed to do a pathology residency in his spare time while in Sioux Falls. Faqian is now a NIH funded Professor of Pathology. XJ told me Faqian was so young when he started med school he was still growing. They were critical in my attracting other competitive cardiovascular researchers, including Rong Liang who has been with me ever since. Rong is another successful MD/PhD researcher from China who was also doing a postdoc with Jeff Robbins group in Cincinnati, where he became good friends with XJ. So, my success story at the CRC in SD was largely based on brilliant Chinese scientists who worked with me or my trainees previously. This connection would also prove critical to my success in NY. Treat your trainees well!

Rong eventually joined my department at NYITCOM. He was initially reticent mainly because he thought I could attract higher quality people in NY than in SD. But, when I showed him that he would have ranked #1 in our last 2 faculty searches, he decided to join me. He has become a lifelong friend for almost 25 years now. Rong is one of the kindest, most humble, and brilliant people I know. He has become one of the most admired faculty at NYITCOM and has been almost continuously funded since our relationship began. His wife, Yuan Huang, MD/MS, also runs an important core facility for our department and provided an earlier photo of isolated myocytes for this book. They are an adorable couple. Rong's Japanese postdoc, Satoru Kobayashi, PhD, was so successful in obtaining major grants that he earned a faculty position at NYITCOM. Satoru's wife, Tamayo, is one of our best technicians. She seems to know everyone's birthday and bakes a cake to celebrate the occasion at work. That's a lot of cakes! Two other Chinese faculty from the basic sciences at USD in Vermillion followed me to NY. Both are cancer researchers. One got his PhD with a Nobel Laureate and did a postdoc at Harvard. The other has been NIH funded since his arrival here. My Russian postdoc, Olga Savinova, from the CRC in Sioux Falls eventually joined me and became our first faculty to obtain a major NIH grant that was submitted from here. My last PhD student from USF in FL is now a pathology professor at NYU in Manhattan. These are only a few of the successful trainees from my labs.

My MD/PhD student from USD, Nate Weltman, also followed me to NY a year after my move. Because institutions invest heavily in these students and do not want to disrupt their education, Nate was able to maintain my lab in SD and continue his research while I remodeled the lab in NY. So, my research had no down time. I was able to remain his supervising professor as an adjunct until he completed the program. Nate managed to get a BS, MD, and PhD without paying any tuition. He had an undergraduate scholarship as a kicker at the University of Virginia and a full scholarship in the MD/PhD program. Nate's PhD was an interesting story. He started his research in my lab in South Dakota, then joined me in NY. He finished his research 6 months early and spent that time in Giorgio Iervasi's lab in Pisa, Italy working on clinical studies related to his animal work in my lab. Giorgio is a good friend who is not only director of one of the largest clinical translational centers in Europe but has published more human studies on thyroid hormone dysfunction in heart failure than anyone. Nate married just before moving to Pisa.

Although his time in Pisa entailed clinical research, it was somewhat of an extended honeymoon filled with many weekend excursions. He published 3 more papers with Giorgio who covered some of his expenses there. Nate did his residency training at Johns Hopkins, considered the world's top medical school.

My brilliant postdoc, Yuefeng Chen (the one accused of leaving a rat hair on the operating table) also joined me. So, about 1/3 of the successful department I built at NYITCOM were people who followed me to NYITCOM from South Dakota. If I had not won the respect and admiration of these wonderful young scientists, my career would likely have taken a more subdued trajectory. It is important for scientists to build productive research groups. You need to be the person with the initial ideas who directs the work. But they do most of the work and often make your ideas blossom and expand into something better than you imagined. When you read about famous scientists, realize that many of their best ideas often come from young people who were training with them. This mentor-mentee interplay is one of the most exciting things about doing science.

This SD contingent served as a springboard to rapidly build a success story at NYITCOM. Another outstanding cardiovascular researcher was my first faculty recruit. Youhua Zhang, MD/PhD was an electrophysiology cardiologist **(EP)** at Fuwai Heart Hospital in Beijing, the top heart hospital in China. He then spent 11 years doing EP research in a famous lab at the Cleveland Clinic before joining us in NY. He runs our small animal physiology core, providing rat and mouse echocardiograms, heart function (left ventricular catheterization), and arrhythmia testing. I would argue that he runs the best small animal cardiovascular physiology core in the US. He discovered an electrical pathway in the heart that is named after him. His wife Ying Li is also an outstanding technician in our department.

In the early days, NYITCOM was a difficult sell. Few PhDs knew anything about DOs at that time. MDs had been working alongside them for many years and had come to know them as clinical equals. While patient understanding lagged, more recently they have grown to know and accept the parity. It is more difficult to obtain a DO than an MD. DOs must know everything that MDs know, plus procedures specific to osteopathy. DO students spend about an extra half day per week learning and practicing what is known as **OMM (Osteopathic Manipulative Medicine)**. While some procedures are not yet backed by experimental proof and need further documentation, many patients are hooked. I personally experienced some of the "magic" after a tennis injury affected an area just above my Achilles tendon. It would not heal until after an OMM manipulation literally allowed me to walk away saying: "What just happened?" Prior to the injury, I could not put any pressure on the front of my foot. I had to walk flat-footed upstairs. This could have been a life changing event for me because I am a passionate walker, 5 or more miles per day since I quit running in my late 50s. I asked Dr Yao, the DO who worked on me, what happened. He said the tendons from the back of the leg extend to the underside of the foot (plantar fascia), wrapping around the bones in the foot (tarsals). Those bones had become misaligned with the injury, and he adjusted them back to normal position. My foot began to heal rapidly. It changed my life.

More recently, DO schools have been trying to experimentally validate some of their most promising procedures. It is challenging because OMM procedures involve physical manipulation and control procedures must also involve some physical manipulation that appears to be similar but of no likely benefit. And, of course, we all know that placebos often show dramatic improvements too. The power of suggestion! I am confident the proof will come for many of these procedures. I am particularly optimistic about OMM procedures that improve lymphatic flow, removing toxins from tissues while stimulating return to a more normal state. I had a few of those procedures and felt wonderful afterwards. Nonetheless, most DOs eventually do only "MD" procedures after training. The time-consuming nature and low reimbursement of OMM procedures may be an impediment.

In addition to a lack of understanding of DOs by PhD biomedical researchers who almost universally train at MD schools, there was very little research infrastructure at NYITCOM when I arrived. There was a small but well-run rodent facility. Kurt Amsler was recruited just prior to me as Associate Dean of Research. Kurt had a broad background in basic and clinical research having worked in both areas at a top school in NYC. Prior to that, Kurt worked at a drug company and understood drug development and patenting. I was very grateful for the groundwork he provided shortly before my arrival. But research deans do not have faculty hiring authority, so it was up to me to build biomedical research at NYITCOM. The process of grant submission and management were also at a very basic level because the needs were minimal. In the early years, it often felt like I was reinventing the wheel at many turns. This was the background for many of my early faculty recruitment interviews, who often came from Ivy League schools. Thank God for my South Dakota contingent, who quickly helped us establish a respectable research appearance. The location was also a big draw, with Atlantic beaches and New York City both ~25 miles away.

DO schools are also different from MD schools in structure. For example, my department (Biomedical Sciences) has physiologists, a microbiologist, a biochemist, a pathologist, a pharmacologist, neuroscientists, and molecular biologists. These are typically independent departments at MD schools, comprising over 8 faculty in each discipline. Notice I stated "a" in some cases. Yes, there is only one faculty in my department for some disciplines when MD schools typically have separate departments for these disciplines with many more faculty. We are also one of the largest medical schools in the US with respect to class size, which is 300 (~1,200 students in all 4 years). On the surface, this seemed like an impossible task to deliver the curriculum with so few faculty. It helps that one can lecture to 300 students as easily as 10. Like many schools, we have moved to an integrated curriculum based on topics, like cardiovascular, renal, pulmonary, for instance, where gross anatomy, histology, physiology, pharmacology, and pathology are taught for those topics in blocks. Like other schools, we continue to tinker with the delivery approach and have a separate group of administrators who coordinate the curriculum. My role as chair was to ensure we have sufficient personnel to teach each area of the basic science curriculum, and our department is responsible for over half of the first two years. This also made faculty recruitment and building research more challenging in the background of teaching needs.

When I arrived, there was only one pathologist at the med school with a recent departure before my arrival. Maria Plummer, MD, was also an early career Assistant Professor. She was assigned to my department, and I needed to fill the open position. I was worried if she could handle the entire pathology curriculum on her own. She assured me she could, and she did. Notably, Maria receives outstanding teacher awards from our students every year. Nonetheless, this became my most pressing recruit. When you have only one teacher for an entire discipline, what do you do if something happens to them?

I quickly found out that pathology is a very difficult recruit and I was also looking for someone who was a good researcher too. We had only one applicant who met both expectations, Alex Vasilyev, MD, PhD, from Harvard. He turned down my initial offer, but I stayed in touch with him. I knew Alex was a pure academic and Harvard overloaded him with clinical duties. We were a good match for him since NYITCOM has no clinical duties in pathology (we are mostly primary care). This meant that he could spend all his time on teaching and research. About a year later, Alex called and asked if the position was still available. I told him I had kept it open for him, although we did not have any other applicants that met my expectations. This time, he accepted. He proved to be an amazing recruit. To give you an idea how focused he was on academia rather than clinical practice, he later told me he was offered a clinical practice position for nearly 4X our salary! This speaks volumes for physicians who see themselves as academics and choose that path despite the financial loss. It also underscores how difficult this recruit was. Thankfully, his wife was also an academic and understood the mentality. Alex was an amazing scientist who later started an innovation and robotics center here. Sadly, he left seven years later due to health concerns.

Fortunately, I got lucky again and recruited Eugen Petcu, MD, PhD. Eugen was in Australia at that time but trained in Boston and wanted to return to the US. He also had family on Long Island. Lucky me! His primary research area is bone regeneration, but he has an international presence in digital pathology, a growing field especially with the innovation of AI. So, not only did I recruit another academic pathologist, but he was also on the cutting edge of a major transformation in pathology. Eugen is licensed to practice in Australia, Romania (his native country), France, Switzerland, Norway, and Massachusetts. So, he had other options. Fortunately, he really liked NYITCOM, the wonderful collegiality, and the productivity of our department. He and several other faculty in the department enjoy going out to happy hour to catch up when we find time.

Like I said earlier, you don't have to be an asshole to be boss. It's not an *ass*et. I have come to realize that one of my biggest strengths is recruiting. Good people like working for me because they know I will have their back if they work hard and are productive. I look for ways to make their life better and make it a point to drop by their offices often and catch up. Sometimes they tell me of an interesting discovery they just had. It is so much fun to share their excitement. I honestly believe that I work for my faculty rather than the other way around. The most difficult challenge for me is to get salary increases for those who are underpaid and highly marketable. I have won a few of those battles. For instance, I got a large pay bump for a female faculty by recruiting a male of the same rank but at a higher salary, which created an inequity. I then highlighted the gender inequity and got a targeted raise for her the next year.

It is also important that faculty know I can make difficult decisions when necessary. I have terminated four faculty since my arrival. These faculty knew they were underperforming and stood out in a department with a growing number of star performers. One found a good position elsewhere. Another was well past retirement age and slipping, literally. She tripped and fell down the stairwell twice. She was ok but seemed annoyed when I pointed to the elevator and suggested she explore that innovative machine. In her case, the dean did not renew her contract. Another was 2 months from age 65 when his contract expired and was not renewed by the dean. I managed to get the dean to extend his contact until he became eligible for Medicare. I remain on good terms with most of them. I had expressed my concerns about their performance years before their departure, and they knew what was needed to keep their jobs. This was outlined in annual evaluations. Another underperforming faculty that I inherited left on her own for another position after some subtle nudging on my part. At the time of writing, 75% of department faculty were recruited by me. This would not have been possible if we had tenure. There is now a push for tenure at the med school since the other schools at NYIT have a tenure program. I am ok with that now because we have only keepers in the department. But the lack of tenure was certainly helpful in building this department so quickly.

I advise my faculty that it is often a good idea to leave their office door open when interacting with members of the opposite sex, particularly when something seems amiss. It's just a good idea. Without going into detail, I have seen both men and women creating issues of concern. Faculty should be cautious of meeting with students of the opposite sex behind closed doors. You never know how desperate a student facing dismissal from med school can become. Of course, faculty members of the opposite sex who know and trust each other, may often need privacy to discuss confidential matters related to students or work that should not be overheard.

I once discovered that employees can also become dangerous when they are dismissed. In one of my previous jobs, our space had a shared secretary with terrible work ethics. She maxed out her vacation and sick leave time and then invented ridiculous stories about having to take off to attend frequent funerals. I began to wonder if she had any living friends left. Some of the stories reminded me of that TV show that always began with a bizarre death (Six Feet Under). One faculty often began the morning with this question: "Is she going to be here today or is there another funeral." The Sanford Research Director eventually fired her. But that was not the end of the story. She often asked me what she could do with her obese, diabetic, meat and potato eating husband, who had heart disease. Of course, we were working in a group of cardiovascular researchers, and these risk factors were often topics of discussion. I told her that portion size is the best target for losing weight. I never mentioned anything about her also being overweight and certainly knew better. After being fired, she told HR that I was harassing her about being overweight, clearly an effort to create grounds for a lawsuit. Desperate people do desperate things. HR investigated and found nothing to support her claim. Her space was in an area shared with several other nearby employees who confirmed her story had no merit. The lesson here is to be careful what you say to employees, particularly troubled employees whose life is clearly a mess.

NYIT was slow to create a new work policy after Covid. During that time, we were pleasantly surprised at faculty productivity while working remotely. Much of our work switched to Zoom for teaching and meetings and that continued post-Covid. The med school finally adopted a post-Covid policy, suggesting faculty spend a minimum of 3 days a week on campus. I generally rate faculty on their accomplishments: publications, funded and submitted grants, teaching evaluations, and service work. If they excel at a high level, I am generally ok with how they do it. That may mean spending more time working from home. Additionally, we have some faculty with over a 2 hour round trip commute. It makes sense for them to work remotely a couple days a week if work can be done in that manner. Of course, where you choose to live and the commuting distance should not dictate attendance at work. The vast majority of faculty working in our department are here every day, particularly those involved in animal and bench research. We are in a new era post-Covid. If someone has outstanding productivity with an exceptional level of quality, I don't see the need to go by their house and see if I can catch them playing soccer with their kids during work hours. I also encourage our employees to take an occasional break and go for a walk in the beautiful forest surrounding our campus. Enjoy the rolling hills, magnificent trees, the old stone bridge, the beautiful pond, and occasional deer and other wildlife.

I inherited a group of neuroscientists when their chair was dismissed. They were publishing well but not in good journals and none had major funding. At that time, few at the med school, particularly clinical faculty, had any idea about the quality of basic science journals and the need to publish in such journals to get grants funded. As a result, reappointment committee members were largely bean counters, counting numbers rather than quality of papers. I urged my neuro faculty to publish in better journals if they wanted to get funding. NIH study section members reviewing their grants look at both quality and number of publications. Neuro faculty responded well and several obtained NIH funding after improving their resumes. Of the original neuro group, 3 remain and they are all keepers. This is the first time in my life to be in a department with no dead weight. That is unusual. There are usually a few tenured "old-timers" in most medical school basic science departments who are performing minimally while counting the days until retirement. Just because you can still fog a mirror does not mean you are qualified for a job in my department.

A good chair (I prefer the neutral term rather than chairman or chairwoman) does not let brilliant faculty members continue to flounder year after year. I saw many examples of this in my career. It is also worth noting that every faculty member is uniquely different. So, different approaches are often necessary. One of my recruits was a brilliant neuroscientist who was intent on publishing only in the world's leading journals. He learned that approach while working at the world's leading research center based on quality of publications (journal Impact Factor). They simply would not allow their researchers to publish in more "pedestrian" journals. The world's leading labs often publish in these high impact journals. But they typically have considerable funding and an army of graduate students and postdoctoral fellows. While the lab may publish a paper of two in journals like *Science* or *Nature* every year, trainees may often leave with few or no publications and a weak springboard to start their careers. Research projects published in these journals are often many years in the works, risky, and with a high rate of failure. I urge my research-intensive faculty to continue their efforts to publish in these elite journals but to avoid putting all their eggs in one basket. They should try to publish a minor paper or more in well-respected journals every year in addition to collaborative studies with others. An example would be the American Journal of Physiology, a journal established in 1898 that is even recognized by the public but has a modest impact factor. In most places, this is the correct recipe for reappointment and promotion, and tenure for those in a tenure system. This brilliant neuroscientist in my department was on the verge of losing his job because he had not published for several years before joining us and was continuing on that trajectory. Yes, he had several papers in high impact journals before joining us, but the clock reset when he took an independent faculty position. After going from a three-year appointment to a one-year contract, a warning that the next contract may be terminal, he finally responded to the road map I laid out. He just got promoted and obtained a fundable score on a large NIH grant. It also looks like he is on the verge of a revolutionary new treatment for a terrible childhood disease. I am so proud of him. I never gave up on him and constantly dropped by his office to check his progress. These are the kind of success stories that really make me happy.

While deans at prestigious allopathic (MD) med schools usually have a strong research background, the opposite is true for DO schools. This is understandable since their primary mission has been medical education. DO schools also require their deans to have a DO. After achieving parity with MD schools, many DO schools are now starting to believe they should contribute to medical knowledge by developing research. I have survived 5 deans since joining NYITCOM. I say "survived" because chairs often feel this way. Increasing resources for departmental research can be a battle with so many departments competing for limited funds. I have found that deans who know nothing about research but think they know everything can sometimes make unfortunate decisions with unfounded confidence. Looking back, I believe Barbara Ross Lee (Dianna Ross' sister) was the best dean I had here regarding research development. She was an outstanding physician and educator and the first black woman to be dean of a US med school (other than a HBCU). She had a limited research background, knew it, and trusted my leadership in developing research at NYITCOM. When things needed to change, she listened while I stated my case. She was also the most fashionable dresser I have ever seen at any med school. She had conservative work outfits, but worthy of a fashion runway. She may have become a world-renowned fashion expert had she not gone into medicine. I never heard her sing or saw her dance, so I can't contribute to that discussion.

The COVID economic crash was particularly hard for higher education institutions, not to mention for students. I am particularly saddened by the experiences of high school and college students during this period. Fortunately, our med school was always one step ahead of the disaster. For instance, we had our faculty trained on Zoom to deliver our curriculum on time with no schedule delays. Other med schools in the area were asking us for advice. NYIT survived this period well thanks to the leadership of our new president Hank Foley. He had to make some hard financial decisions, but we delivered for our students. Before Covid, we had division II sports teams in basketball, lacrosse, soccer, women's softball, and a division I baseball team for men. We are a tech school with many non-traditional students and about half of our enrollment being graduate, rather than undergraduate. Attendance at sporting events was almost non-existent. So, the president decided to close all these sports programs. We weathered the storm well. One reason European universities are less expensive to run is likely due to the savings from not having expensive sports programs. Keeping up with the Joneses for US universities is very expensive with so many men's and women's sports to support.

Our departmental research success quickly spread throughout the campus. There was a major uptick in engineering and life sciences research, in particular. They were also obtaining grants from the NIH and other sources. The recruitment of a world-class researcher, Hank Foley, as president of NYIT led to a dramatic investment in research. A new research building opened in the summer of 2025 at the medical school. An imaging center is at the heart of this Biomedical Research Innovation and Imaging Center (BRIIC). Several years earlier, I had recruited an outstanding imaging expert from Albert Einstein School of Medicine, Randy Stout, a neuroscientist, who became the BRIIC Director. The imaging center probably has the most advanced array of biological imaging microscopes on Long Island. It also has a fMRI (functional Magnetic Resonance Imaging) facility to expand OMM clinical research. When I came to NYIT, no one ever dreamed it would lead to this level of research success. It did not happen overnight. It took years of focus by a hard-working team of wonderful people.

NYIT has campuses on Long Island, Manhattan (at Columbus Circle near Central Park), and a clinical campus in Jonesboro, AR. The additional clinical site, New York Institute of Technology College of Osteopathic Medicine at Arkansas State is a heartwarming success story located on the Arkansas State University campus in Jonesboro. NYITCOM was approached by Arkansas to help solve the health care crisis in the Mississippi Delta region. Barbara Ross Lee was the major driving force for development of this campus and became the first dean. There was major support from the AR governor who realized that opening a DO med school would help provide improved health care to the region at little cost to the state since we are a private school. The first graduates from our AR med school are now completing residencies, starting clinical practice, and achieving the mission of providing health care to poor areas in the Delta region. It is an interesting pairing of a liberal Northeast state with a conservative Southern state. But we made it work, thanks to good leadership.

After announcing my retirement when my contract expires in 2027, I stepped down as chair. We have an excellent acting chair, hand-picked by me. These are very tough times for biomedical research across the US with severe chaos and dismissal of world class investigators and administrators at NIH starting in 2025. I am grateful that these adverse changes came *after* I had already built a major research success story here. It will be easier to hold our ground during these devastating times for biomedical research where research expansion is likely on hold across the US. AI is also having a major impact on universities and the public. I am optimistic that we will weather these storms. For me, this looked like a good time to write a book and prepare for retirement.

Epilogue

At age 74, I am finally accepting the fact that I will not live forever and should travel the world while I am still in great health. Neither of my daughters are married yet, but I am looking forward to seeing grandchildren at some point. I have been wanting to write this book for a while but simply could not find time while running the largest department at our medical school. I am pleasantly surprised by how much fun the book has been to write and how easily it flowed from my head. Regarding my scientific career, I am optimistic that my life's work will finally translate to improved patient outcomes. While the establishment remains focused on the next blockbuster drug and all the associated profits, I believe the low T3-heart failure story is a fundamental truth in medicine. I believe early identification and treatment of this condition will become standard medical practice and lead to longer survival and improved quality of life for millions of people around the world going forward. I just hope it happens before I am gone. I am not alone in this endeavor. It is catching on around the world. The snowball is rolling down the mountain and gaining speed.

I hoped to accomplish many things with this book. Most importantly, I wanted students interested in science but struggling with what they will do with their life to consider taking the path I took. I wanted to let them walk in my shoes and see if they fit. I am just a regular guy who loves science and decided to pursue it. I have been good at creating and identifying opportunities in my career and surrounding myself with amazing people to carry out my visions. Perhaps a few readers will realize that jumping on a golden opportunity can have a major impact on your life trajectory, and then do it! I hope that readers learned something about heart failure since so many families are affected by this killer. Readers may also have learned a little about the inside workings of medical research, medical schools, and the associated politics. Of course, this is just one snapshot from the perspective of one person. We all have a uniquely different story. Hopefully, you enjoyed my quirky sense of humor. I have always been a playful kid at heart. I have had an amazing life and am not done yet.

While writing this book, I got some interesting news from the University of South Dakota School of Medicine. My friend, Dr Tim Ridgeway, has become the new dean. My oldest daughter and his daughter were friends while growing up and our families spent some time together. Tim is a wonderful, considerate, down-to-earth guy. Guess what he just announced? He is consolidating the medical school in Sioux Falls! Over the next few years, the basic sciences in Vermillion will be moving there. He gave several reasons for the move. Foremost, it was for the good of the students. Not only will this move allow better integration of clinical and basic science education, but students will also have many more research opportunities, especially in basic research. Promoting this idea is what got me into trouble with the previous USD president, who retired several years ago. I have always fought for what is right. It has not been an easy path, but I stuck to my guns. I came close to disaster on few occasions while fighting against racial discrimination and standing up to bad leadership decisions. It is not a pathway for the faint of heart. But success is extremely gratifying. Those who just go with the status quo are often forgotten. You live one life. Do you really want to be forgotten? Maybe a scientist will read one of my papers 100 years from now and wonder what kind of person I was. I sometimes do that when reading innovative papers written long ago. I enjoy reading biographies about scientists too. Most recently, I have read books about Galileo, Tony Fauci, and Carl Sagan. But I also read a book about Dolly Parton and how she develops her wardrobe. I guess I just like interesting people.

Glossary Of Terms

ACE inhibitor: Angiotensin Converting Enzyme inhibitor

AVF: Arteriovenous Fistula, abnormal connection between an artery and vein, bypassing the capillary bed

BNP: Brain Natriuretic Peptide, a common biomarker used in heart failure

CO: Cardiac Output, the volume of blood the heart pumps per minute

COBRE: Center of Biomedical Research Excellence

Concentric hypertrophy: thickening of the ventricular wall without chamber dilatation due to hypertension

CRC: Cardiovascular Research Center

DCM: idiopathic **D**ilated **C**ardio-**M**yopathy, unknown cause, but not ischemic

Eccentric hypertrophy: enlargement of the heart from volume overloading leading to a proportional increase in chamber diameter and wall thickness

EP: electrophysiologist, cardiologist specializing in electrical activity of the heart

HBCU: Historically Black Colleges and Universities

Heart failure: a syndrome and common end point for heart diseases. It is characterized by a dilated, dysfunctional heart and the likelihood of an earlier death

HFpEF: Heart Failure with preserved Ejection Fraction, non-dilated form

HFrEF: Heart Failure with reduced Ejection Fraction, dilated form

IACUC: Institutional Animal Care and Use Committee

INBRE: IDeA Networks of Biomedical Research Excellence

IRB: Institutional Review Board, approves human research protocols

Low T3 Syndrome: normal T4, normal TSH, but low T3

MBF: Microvascular blood flow

MI: Myocardial Infarction

Morphometry: quantitative morphology or measuring structural features

mRNA: Messenger RNA, message from gene that encodes proteins

MS: Mass Spectrometry, an analytical technique used to identify compounds

NYITCOM: New York Institute of Technology, College of Osteopathic Medicine

OMM: Osteopathic Manipulative Medicine

Phenotype: an observable trait that can be seen or measured

PPGs: Program Project Grants

RAAS: Renin-Angiotensin-Aldosterone System, regulates blood pressure and blood volume

rT3: reverse T3, inactive form of thyroid hormone

SDSU: South Dakota State University

SHR: Spontaneously Hypertensive Rat, a genetic model of hypertension.

SR: sarcoplasmic reticulum, intracellular organelle in muscle cells

STEM: Science, technology, engineering, mathematics

Subclinical Hypothyroidism: normal T3 and T4, but elevated TSH.

Synthroid: man-made form of T4

T3: tri-iodo thyronine, more active thyroid hormone

T4: tetra-iodo thyronine, prohormone that becomes activated to T3 after iodine removal

TRH: Thyroid Releasing Hormone, secreted by the hypothalamus

TSH: Thyroid Stimulating Hormone, secreted by the pituitary gland

UAB: University Alabama Birmingham

USD: University of South Dakota

USF: University of South Florida

References

1. **Gerdes AM, and Kasten FH**. Morphometric study of endomyocardium and epimyocardium of the left ventricle in adult dogs. *Am J Anat* 159: 389-394, 1980.

2. **Weltman NY, Wang D, Redetzke RA, and Gerdes AM**. Longstanding hyperthyroidism is associated with normal or enhanced intrinsic cardiomyocyte function despite decline in global cardiac function. *PLoS One* 7: e46655, 2012.

3. **Gerdes AM, Kellerman SE, Moore JA, Muffly KE, Clark LC, Reaves PY, Malec KB, McKeown PP, and Schocken DD**. Structural remodeling of cardiac myocytes in patients with ischemic cardiomyopathy. *Circulation* 86: 426-430, 1992.

4. **Tamura T, Onodera T, Said S, and Gerdes AM**. Correlation of myocyte lengthening to chamber dilation in the spontaneously hypertensive heart failure (SHHF) rat. *J Mol Cell Cardiol* 30: 2175-2181, 1998.

5. **Gerdes AM, Moore JA, Hines JM, Kirkland P, and Bishop SP**. Regional differences in myocyte size in normal rat heart. *Anat Rec* 215: 420-426, 1986.

6. **Grossman W, Jones D, and McLaurin LP**. Wall stress and patterns of hypertrophy in the human left ventricle. *J Clin Invest* 56: 56-64, 1975.

7. **Bai SL, Campbell SE, Moore JA, Morales MC, and Gerdes AM**. Influence of age, growth, and sex on cardiac myocyte size and number in rats. *Anat Rec* 226: 207-212, 1990.

8. **Campbell SE, and Gerdes AM**. Regional changes in myocyte size during the reversal of thyroid-induced cardiac hypertrophy. *J Mol Cell Cardiol* 20: 379-387, 1988.

9. **Liu Z, and Gerdes AM**. Influence of hypothyroidism and the reversal of hypothyroidism on hemodynamics and cell size in the adult rat heart. *J Mol Cell Cardiol* 22: 1339-1348, 1990.

10. **Tang YD, Kuzman JA, Said S, Anderson BE, Wang X, and Gerdes AM**. Low thyroid function leads to cardiac atrophy with chamber dilatation, impaired myocardial blood flow, loss of arterioles, and severe systolic dysfunction. *Circulation* 112: 3122-3130, 2005.

11. **Zierhut W, Zimmer HG, and Gerdes AM**. Effect of angiotensin converting enzyme inhibition on pressure-induced left ventricular hypertrophy in rats. *Circ Res* 69: 609-617, 1991.

12. **Liu Z, Hilbelink DR, Crockett WB, and Gerdes AM**. Regional changes in hemodynamics and cardiac myocyte size in rats with aortocaval fistulas. 1. Developing and established hypertrophy. *Circ Res* 69: 52-58, 1991.

13. **Gerdes AM**. Cardiac myocyte remodeling in hypertrophy and progression to failure. *J Card Fail* 8: S264-268, 2002.

14. **Gerdes AM, Clark LC, and Capasso JM**. Regression of cardiac hypertrophy after closing an aortocaval fistula in rats. *Am J Physiol* 268: H2345-2351, 1995.

15. **Pfeffer JM, Pfeffer MA, Mirsky I, and Braunwald E**. Regression of left ventricular hypertrophy and prevention of left ventricular dysfunction by captopril in the spontaneously hypertensive rat. *Proc Natl Acad Sci U S A* 79: 3310-3314, 1982.

16. **Pfeffer MA, and Braunwald E**. Ventricular remodeling after myocardial infarction. Experimental observations and clinical implications. *Circulation* 81: 1161-1172, 1990.

17. **Pfeffer MA, Braunwald E, Moye LA, Basta L, Brown EJ, Jr., Cuddy TE, Davis BR, Geltman EM, Goldman S, Flaker GC, and et al.** Effect of captopril on mortality and morbidity in patients with left ventricular dysfunction after myocardial infarction. Results of the survival and ventricular enlargement trial. The SAVE Investigators. *N Engl J Med* 327: 669-677, 1992.

18. **Zimmer HG, Gerdes AM, Lortet S, and Mall G**. Changes in heart function and cardiac cell size in rats with chronic myocardial infarction. *J Mol Cell Cardiol* 22: 1231-1243, 1990.

19. **Kuo PL, Lee H, Bray MA, Geisse NA, Huang YT, Adams WJ, Sheehy SP, and Parker KK**. Myocyte shape regulates lateral registry of sarcomeres and contractility. *Am J Pathol* 181: 2030-2037, 2012.

20. **Tamura T, Said S, Harris J, Lu W, and Gerdes AM**. Reverse remodeling of cardiac myocyte hypertrophy in hypertension and failure by targeting of the renin-angiotensin system. *Circulation* 102: 253-259, 2000.

21. **Owan TE, Hodge DO, Herges RM, Jacobsen SJ, Roger VL, and Redfield MM**. Trends in prevalence and outcome of heart failure with preserved ejection fraction. *N Engl J Med* 355: 251-259, 2006.

22. **Chien KR, Frisen J, Fritsche-Danielson R, Melton DA, Murry CE, and Weissman IL**. Regenerating the field of cardiovascular cell therapy. *Nat Biotechnol* 37: 232-237, 2019.

23. **Akirov A, Fazelzad R, Ezzat S, Thabane L, and Sawka AM**. A Systematic Review and Meta-Analysis of Patient Preferences for Combination Thyroid Hormone Treatment for Hypothyroidism. *Front Endocrinol (Lausanne)* 10: 477, 2019.

24. **Vargas-Uricoechea H, and Wartofsky L**. LT4/LT3 Combination Therapy vs. Monotherapy with LT4 for Persistent Symptoms of Hypothyroidism: A Systematic Review. *Int J Mol Sci* 25: 2024.

25. **Morkin E**. Regulation of myosin heavy chain genes in the heart. *Circulation* 87: 1451-1460, 1993.

26. **Schwartz K, and Mercadier JJ**. Molecular and cellular biology of heart failure. *Curr Opin Cardiol* 11: 227-236, 1996.

27. **Lowes BD, Minobe W, Abraham WT, Rizeq MN, Bohlmeyer TJ, Quaife RA, Roden RL, Dutcher DL, Robertson AD, Voelkel NF, Badesch DB,**

Groves BM, Gilbert EM, and Bristow MR. Changes in gene expression in the intact human heart. Downregulation of alpha-myosin heavy chain in hypertrophied, failing ventricular myocardium. *J Clin Invest* 100: 2315-2324, 1997.

28. **Gerdes AM**. Cardiac remodeling in heart disease and the importance of thyroid hormones. *Am J Physiol Heart Circ Physiol 328:H1005-H1018,* 2025.

29. **Chen YF, Weltman NY, Li X, Youmans S, Krause D, and Gerdes AM**. Improvement of left ventricular remodeling after myocardial infarction with eight weeks L-thyroxine treatment in rats. *J Transl Med* 11: 40, 2013.

30. **Weltman NY, Pol CJ, Zhang Y, Wang Y, Koder A, Raza S, Zucchi R, Saba A, Colligiani D, and Gerdes AM**. Long-term physiological T3 supplementation in hypertensive heart disease in rats. *Am J Physiol Heart Circ Physiol* 309: H1059-1065, 2015.

31. **Weltman NY, Ojamaa K, Schlenker EH, Chen YF, Zucchi R, Saba A, Colligiani D, Rajagopalan V, Pol CJ, and Gerdes AM**. Low-dose T(3) replacement restores depressed cardiac T(3) levels, preserves coronary microvasculature and attenuates cardiac dysfunction in experimental diabetes mellitus. *Mol Med* 20: 302-312, 2014.

32. **Sato Y, Yoshihisa A, Kimishima Y, Kiko T, Kanno Y, Yokokawa T, Abe S, Misaka T, Sato T, Oikawa M, Kobayashi A, Yamaki T, Kunii H, Nakazato K, and Takeishi Y**. Low T3 Syndrome Is Associated With High Mortality in Hospitalized Patients With Heart Failure. *J Card Fail* 25: 195-203, 2019.

33. **She J, Feng J, Deng Y, Sun L, Wu Y, Guo M, Liang X, Li J, Xia Y, and Yuan Z**. Correlation of Triiodothyronine Level with In-Hospital Cardiac Function and Long-Term Prognosis in Patients with Acute Myocardial Infarction. *Dis Markers* 2018: 5236267, 2018.

34. **Pol CJ, Muller A, Zuidwijk MJ, van Deel ED, Kaptein E, Saba A, Marchini M, Zucchi R, Visser TJ, Paulus WJ, Duncker DJ, and Simonides WS**. Left-ventricular remodeling after myocardial infarction is associated with a cardiomyocyte-specific hypothyroid condition. *Endocrinology* 152: 669-679, 2011.

35. **Weltman NY, Ojamaa K, Savinova OV, Chen YF, Schlenker EH, Zucchi R, Saba A, Colligiani D, Pol CJ, and Gerdes AM**. Restoration of cardiac tissue thyroid hormone status in experimental hypothyroidism: a dose-response study in female rats. *Endocrinology* 154: 2542-2552, 2013.

36. **Gil-Cayuela C, Ortega A, Tarazon E, Martinez-Dolz L, Cinca J, Gonzalez-Juanatey JR, Lago F, Rosello-Lleti E, Rivera M, and Portoles M**. Myocardium of patients with dilated cardiomyopathy presents altered expression of genes involved in thyroid hormone biosynthesis. *PLoS One* 13: e0190987, 2018.

37. **Gil-Cayuela C, Rosello LE, Tarazon E, Ortega A, Sandoval J, Martinez-Dolz L, Cinca J, Jorge E, Gonzalez-Juanatey JR, Lago F, Rivera M, and Portoles M**. Thyroid hormone biosynthesis machinery is altered in the ischemic myocardium: An epigenomic study. *Int J Cardiol* 243: 27-33, 2017.

38. **Rajagopalan V, Gorecki M, Costello C, Schultz E, Zhang Y, and Gerdes AM**. Cardioprotection by triiodothyronine following caloric restriction via long noncoding RNAs. *Biomed Pharmacother* 131: 110657, 2020.

39. **Del Buono MG, Montone RA, Camilli M, Carbone S, Narula J, Lavie CJ, Niccoli G, and Crea F**. Coronary Microvascular Dysfunction Across the Spectrum of Cardiovascular Diseases: JACC State-of-the-Art Review. *J Am Coll Cardiol* 78: 1352-1371, 2021.

40. **Bianco AC, and Casula S**. Thyroid hormone replacement therapy: three 'simple' questions, complex answers. *Eur Thyroid J* 1: 88-98, 2012.

41. **Amin A, Chitsazan M, Taghavi S, and Ardeshiri M**. Effects of triiodothyronine replacement therapy in patients with chronic stable heart failure and low-triiodothyronine syndrome: a randomized, double-blind, placebo-controlled study. *ESC Heart Fail* 2: 5-11, 2015.

42. **Bunevicius R, Varoneckas G, Prange AJ, Jr., Hinderliter AL, Gintauskiene V, and Girdler SS**. Depression and thyroid axis function in coronary artery disease: impact of cardiac impairment and gender. *Clin Cardiol* 29: 170-174, 2006.

43. **Mayer O, Jr., Simon J, Cech J, Rosolova H, Hrbkova J, Pikner R, and Trefil L**. Even mild changes in free thyroxine could influence the degree of heart failure measured by its biological surrogates. *Physiol Res* 57: 525-529, 2008.

44. **Pfister R, Strack N, Wielckens K, Malchau G, Erdmann E, and Schneider CA**. The relationship and prognostic impact of low-T3 syndrome and NT-pro-BNP in cardiovascular patients. *Int J Cardiol* 144: 187-190, 2010.

45. **Du JB, Da CH, Zhao Y, Guo Y, Guo G, Ju TF, and Xu YP**. The role of brain natriuretic peptide and serum triiodothyronine in the diagnosis and prognosis of chronic heart failure. *Acta Cardiol* 67: 291-296, 2012.

46. **Cameron VA, and Ellmers LJ**. Minireview: natriuretic peptides during development of the fetal heart and circulation. *Endocrinology* 144: 2191-2194, 2003.

47. **Wang K, Ojamaa K, Samuels A, Gilani N, Zhang K, An S, Zhang Y, Tang YD, Askari B, and Gerdes AM**. BNP as a New Biomarker of Cardiac Thyroid Hormone Function. *Front Physiol* 11: 729, 2020.

48. **Zhang H, Che W, Shi K, Huang Y, Xu C, Fei M, Fan X, Zhang J, Hu X, Hu F, Qin S, Zhang X, Huang Q, and Yu F**. FT4/FT3 ratio: A novel biomarker predicts coronary microvascular dysfunction (CMD) in euthyroid INOCA patients. *Front Endocrinol (Lausanne)* 13: 1021326, 2022.

49. **Kountz WB**. Vascular degeneration in hypothyroidism. *AMA Arch Pathol* 50: 765-777, 1950.

50. **An S, Gilani N, Huang Y, Muncan A, Zhang Y, Tang YD, Gerdes AM, and Ojamaa K**. Adverse transverse-tubule remodeling in a rat model of heart failure is attenuated with low-dose triiodothyronine treatment. *Mol Med* 25: 53, 2019.

51. **Charest A, Nasta N, Siddiqui S, Menkes S, Thomas A, Saad D, Forman J, Huang X, Sison CP, Gerdes AM, Stout RF, and Ojamaa K**. Nanoscale

organization of cardiac calcium channels is dependent on thyroid hormone status. *Am J Physiol Heart Circ Physiol* 327: H1309-H1326, 2024.

About the Book

Over the years, the author has become increasingly concerned about the lack of bright young people in the US embarking on a career in scientific research. The infusion of foreign scientists who kept us afloat over the past few decades appears to have ended. Indeed, America is at a critical juncture where we are unprepared to continue our dominance of science while the rest of the world is on a fast pace to catch or surpass us. A major driving force for writing this book was to let young people who love science know what it is like to be a scientist. It is a lot more exciting and rewarding than most people could ever imagine. Gerdes challenges readers to walk through life in his shoes while daring them to use the word boring. You may never look at scientists the same way again.

www.ingramcontent.com/pod-product-compliance
Lightning Source LLC
LaVergne TN
LVHW050816061126
851857LV00012B/826
9798240832369